Alberto Palazzi

Relativity from Lorentz to Einstein

A Guide for Beginners, Perplexed and Experimental Scientists

GogLiB

ISBN: 9788897527411
First edition: January 2018 (F)

Contents

Contents

Contents

Foreword

There are thousands of books that expose and make popular Einstein's relativity theory with different levels of complexity. This literature has developed since 1919, when the theory of general relativity unexpectedly exploded in popularity worldwide, uninterrupted to this day. Within a couple of years, hundreds of popularizations[1] were published concerning both the theory of special relativity, which dates back to 1905, and that of general relativity, developed by Einstein in the following ten years. The production of literature continued uninterrupted always exhibiting the same theory with the same arguments and the same expressions. The standard way of describing special relativity was defined by different authors at a time roughly between 1911 and 1925, and has been maintained to date. If at a distance of so many years new books are written, that are very similar to the precedents, there must be a problem. In this discussion we adopt a different perspective in order to highlight the exact meaning of the theory of special relativity, to allow the reader to judge in full the kinematic part of theory, i.e. the one that speaks of space, time, and clocks, and to help the reader understand how the formula $E=mc^2$ can be derived.

In relativistic literature we find three kinds of books: introductory, popular, and specialist. In each kind there are many versions that are similar among them: all books converge towards the same results and all proceed with the same criterion. Those who have consulted many of them recognize familiar issues in each book and look for a single page or a single observation, hoping to shed light on a controversial point and to find help in grasping the whole of the theory. This is a common experience because many people have tried to study relativity and have grasped the pattern of reasoning and the elements of theory, but are still perplexed about the overall connection between them. However, especially with regard to special relativity, the whole literature can be divided into three sets of introductory, popular and specialized books. In the group of introductory treatises there are the simplest ones, which are merely descriptive and serve to give a first idea of the theory, but they can neither explain nor make anything understood. They only serve to create expectations and

---

[1] Isaacson 2007, Chapter 12.

provoke curiosity in the reader, making the conclusions of the theory known but leaving them unexplained. The books of this group are all superfluous, because what they can teach is already known through movies, television broadcasts, encyclopaedia entries and newspaper articles. In short, they repeat commonplaces and common knowledge.

The second group of treatises develops the reasoning process of the theory by using simple terms and a little algebra, but when such books are well written, they are able to give the readers an exact knowledge of the theory. This is because special relativity, also in its discussion by Einstein, makes almost no use of superior maths, except in the electrodynamic part where this is indispensable.

The third group of theory exposures is aimed at university education and at the specialist public and therefore it cannot be used by the reader who approaches it without specific training in physics, mathematics or engineering. However, this limit is far less significant than what one would think for readers who are not able to go beyond the level of the books of the second group, because the main issue of special relativity is not mathematics: it is logic. There is a logical process to follow, which is not easy, but the necessary math is the one of the school, not of the university.

This takes us to the heart of the problem: the whole relativistic literature has the defect of not fully explaining the premises and assumptions that must be made to understand the deduction of the consequences, and in this fault lies the great difficulty that almost everyone experiences in the theory. Indeed, it has often been found, and teachers confirm it, that the smartest students, those who want to go deep into things, are those who find more difficulty in understanding the deep connection between the elements of special relativity, while those who are content to learn how to repeat certain judgments and how to apply some not particularly complex mathematical formulas are usually satisfied both by the theory and by themselves. Before beginning the study of relativity, those who know just the most general topics will probably start from the view that the difficulty of theory lies in the acceptance of the non-Euclidean description of space-time, going against habits consolidated for centuries. Given what is known before acquiring accurate knowledge of the theory this suspicion is reasonable and justified. But by carefully looking at the details we shall see that the concepts of the theory present no difficulty in themselves, and that the lack of understanding depends on a completely different and unexpected factor.

This book was written for two kinds of readers: those who know all the famous images and suggestions associated with special relativity and are curious about it, but who know nothing, except for having heard something or read some article or introductory book, and those who know the fundamentals, having studied them at the popular or specialist level, but are not convinced that they have understood and assimilated the concatenation of the elements of the theory. Many readers and scholars are in this condition even after going deep into the technical elements of the theory: they know the theory of Lorentz's transformations, Minkowski geometry, the derivation of the "World's most renowned formula", $E=mc^2$, from the space-time theory, but they recognize their tenuous grasp of the connection between the parts of the theory and the consistency of the demonstrations.

So this book describes the theory of special relativity with the simplicity necessary for it to be understood by non-specialist readers, who are only required to read it carefully. At the same time it analyzes the logic implied by the typical arguments of relativistic literature, and sheds light on the premises that for long tradition are not expressed with the necessary clarity, and therefore are at the root of the difficulty of understanding the theory. This book was written taking into account both the beginner's and the perplexed people's needs, the needs of those who know the elements of special relativity but feel they have not understood it. Therefore, readers already familiar with fundamentals of the theory will recognize many concepts they already understood and will read quickly or skip the elementary expositions of the things they already know, to dwell on the discussions of the difficulties.

Obviously, this book was not written for those who know they have understood and assimilated the theory and do not feel the need for any clarification.

This book is, in a certain sense, a practical manual that gives the readers tips to guide them into a subject about which it cannot be the only source. As the readers assimilate this book, they will feel the need to compare it with relativistic literature, in which at that point they will feel comfortable. As for the reading method, note that in this book you will find few figures and few formulas. Better read it holding paper and pencil at hand to write down formulas and draw a few figures following the text instructions. Then, since relativistic literature is huge and full of repetitions of the same things, I have chosen to minimize the space dedicated to the most famous topics. This book should serve as an introductory guide to reading any other source about special relativity, of medium or specialist level. You can read other sources right away after finishing this, or even at the same time, by going to the specialist discussion of the single arguments after assimilating the introduction to the background ideas you find clearly stated here in a way that you will not find elsewhere. In particular, I advise to read, in parallel with this book, the popular account of the theory written by Einstein himself in 1916, widely available in many economic editions[2].

With regard to the logical structure of the deductive process of special relativity, there are many things yet to be said and therefore this book devotes the least possible space to dealing with what is known or is explained extensively elsewhere. This book explicitly states all that is necessary to know before entering into the deductive mechanism of special relativity, but for the simplest and most common notions, mostly historical and factual data on which there are no difficulties or discussions, you will often find the indication "see Encyclopaedia". It means that in order to deepen what may not be known or enough clear you are advised to stop the reading of this book and take a look at the subject matter in an encyclopaedia. Even Wikipedia will be fine, but with the warning to prioritize the entries in English rather than those in other languages.

Finally, for those who want to go back to school physics before getting into the study of special relativity, I recommend a basic English textbook, *Understand Physics: Teach Yourself* by Jim Breithaupt, a quick, pleasant, and effective read. I also recommend reading the book written and published by Einstein in 1938 in collaboration with Leopold Infeld: *The Evolution of Physics: The*

---

[2] Einstein, 1916.

Growth of Ideas from Early Concepts to Relativity and Quanta. This book by Einstein introduces the issue of relativity with total clarity as far as the antecedents are concerned. But as it moves to special relativity, it proposes reflections and considerations on it that are relevant to those who already know the theory, without being able to be a useful textbook for those who come to it for the first time. It is therefore advisable to read the first and second part, where the preconditions of special relativity are exposed, before going to study, and the third and fourth part after acquiring knowledge of the theory. The book of Einstein and Infeld is complete, it also deals with general relativity and refers to quantum physics from the point of view of Einstein, who, as is well known, was not favourable toward it.

In general, the biography of the author of a theory is not relevant for the judgment of her or his work; and as a rule, one should concentrate on studying a theory by remaining ignorant of the biography of the author, in order to avoid giving trivial psychological labels to what is not understood. But in Einstein's case, it is imperative to keep in mind the context of the theory, and therefore also the biography, for which I refer to a recent and somewhat voluminous book, *Einstein. His Life and Universe* by Walter Isaacson. We shall see that in the case of Einstein the history of the theory is particularly relevant for understanding the history of its fame and its acceptance by the mainstream of the scientific community. The theory of special relativity was born to answer a very specific question that in most of the books is completely ignored, and in some other treatise is relegated to the background. We shall note that this fact undermines the understanding of the deductive chain of the theory, but we shall also understand why this change of perspective occurred, which at first sight should appear curious. The theory of special relativity quickly assumed a very different meaning than the original purpose, and we shall see why.

Some final practical information: the footnotes in the text contain only references to the bibliography, so you do not have to read them if you do not want to trace the sources of information. Any marginal remarks are not in footnotes, but are incorporated in the text. In the English translation of this book the concept of velocity has been referred to with the simple word "speed" instead of "velocity", that means a vector speed having a given direction and sense. The popular character of this book allows this simplification. Only in literal quotations of English texts the term "velocity" will occur.

1. *Basics*

1.1 *Special relativity and general relativity*

I keep the promise made in the introduction and refer the reader to the Encyclopaedia (any Encyclopaedia, or any other source about the subject) for the preliminary historical notions. I just mention here what is indispensable to know.

Albert Einstein was born in 1879 in Ulm, in southern Germany, and because of family vagaries and vicissitudes, he lived in Italy and Switzerland the years of his youth. Legend has it that he was a bad student, but the study of his biography reveals that he was sometimes a brilliant student, but always rebellious and inclined to contradict teachers. It is interesting to note that, since his father and uncle were pioneer entrepreneurs in the field of electrical installations, Einstein was familiar with the properties of electrical equipment since he was young and could understand their peculiarities with great and innate ease. At sixteen he wrote an essay, conventional in content, about ether and electromagnetic theory, about what would be the central issue of special relativity, and at the same age he was able to help adults in calculations for the sizing up of electrical equipment[3]:

> Einstein spent the spring and summer of 1895 living with his parents in their Pavia apartment and helping at the family firm. In the process, he was able to get a good feel for the workings of magnets, coils, and generated electricity. Einstein's work impressed his family. On one occasion, Uncle Jakob was having problems with some calculations for a new machine, so Einstein went to work on it. "After my assistant engineer and I had been racking our brain for days, that young sprig had got the whole thing in just fifteen minutes," Jakob reported to a friend. "You will hear of him yet."

Einstein got a diploma with a mediocre score, because of controversies with his teachers at the Federal Polytechnic of Zurich, and it was not easy for him to find an occupation. His temper led him to experience intellectual unemployment, at that time more rare than today. Finally, he found a modest job, which anyway was paid about twice the usual salary, at the Swiss Federal Bureau of Patents and in

---

[3] Isaacson 2007, Chapter 2.

1905 he published four papers in a major journal, *Annalen der Physik*, directed by Max Planck. Of the four papers the fourth is an appendix of the third and is only three pages long.

The year 1905, for this reason, is known as Einstein's "annus mirabilis", and the four papers have the titles:

- "On a heuristic point of view concerning the production and transformation of light"[4], on the photoelectric effect. This paper is not about relativity, but we shall refer to its content later on, because we shall use the concept of photon, or "electromagnetic quantum". This essay (and not the theory of relativity) later was the motivation of the Nobel Prize conferred on Einstein in 1921. To understand special relativity, it is useful, though not strictly necessary, to have knowledge of the notion of photon, of which there is a clear explanation in Chapter 8 of Breithaupt's book.

- "On the movement of small particles suspended in stationary liquids required by the molecular-kinetic theory of heat"[5], a paper concerning a particular phenomenon, Brownian motion (see Encyclopaedia), from which Einstein draws important consequences for atomic theory. This paper is not concerned with relativity and we shall not deal with it.

- "On the electrodynamics of moving bodies"[6], the paper that is the primary source of special relativity, and which we shall always refer to with the abbreviation ED1905. Note that the title does not correspond to the idea that we have about the contents of special relativity: there is nothing that seems to concern space, time and clocks.

- "Does the inertia of a body depend upon its energy content?"[7], a three-page appendix to the previous paper on electrodynamics, where is developed what was hinted in the main paper about the equivalence between mass and energy. It is easy to understand that here we shall find the famous equation expressing the amount

---

[4] *Annalen der Physik*, 17, 132-148.

[5] *Annalen der Physik*, 17, 549–560.

[6] *Annalen der Physik*, 17, 891–921.

[7] *Annalen der Physik*, 18, 639–641.

of energy released as a result of mass loss in a chemical transformation and in a nuclear reaction.

So in 1905, through the paper "On the Electrodynamics of moving bodies" and the appendix "Does the inertia of a body depend upon its energy content?" the theory of special relativity was published. It includes these three elements:

- Einstein's completely unedited theory of space and time with respect to systems in uniform rectilinear motion;

- the explicit expression of a new conception of the relationship between mass and energy, which was emerging from others' research of that time;

- a proposed solution to a very specific theoretical problem related to the theory of electromagnetic phenomena. This latter issue gives the title to the research, but it is the least known of all in the common cognition of special relativity. We shall see that this omission is the fundamental cause of the misunderstanding of the whole theory.

Special relativity does not require superior math tools, but only algebraic ones. It is limited to the condition of uniform rectilinear motion, and to overcome this limitation in the ten years after 1905 Einstein developed the theory of general relativity, which concerns accelerated motion and gravity. The theory of general relativity uses superior, highly sophisticated mathematical tools, and we shall not deal with it at all, not only because of the greater difficulty, but because one cannot commit a greater error than dealing with general relativity before focusing on special relativity problems and its specific issues. Without fully understanding special relativity, the study of general relativity may only be mnemonic, just a superficial knowledge of the conclusions of the theory.

Both relativities were described by Einstein in papers that make up their fundamental text: we have already encountered those relating to special relativity, those on general relativity were published in 1915 and 1917 (see Encyclopaedia). Of both theories Einstein wrote in 1916 a version he considered popular, which is a useful and highly advisable reading, but that is not easy and not exactly popular, and is often quoted by relativistic literature as an important source.

1.2   Space measures: the sample ruler problem

Before Einstein, scientists had reached the awareness that human knowledge cannot come to absolutely valid determinations of the unit of measure of space, the unit of measure of time, and the attributions of simultaneity to events. At the dawn of the twentieth century this subject was particularly in vogue, because at that time Henri Bergson's philosophy was spreading and becoming popular: this philosopher reflected widely on the difference and the relationship between the concrete time actually experienced by the consciousness (which lives in the present, and there accumulates memories and mixes them with expectations of the future) and the abstract time of the scientific object construction processes[8]. Henri Poincaré's methodological studies[9] were also widely known: he was an important physicist, a specialist in measure-related problems, and for a long time he was also an active member of international institutions for the unification of measurement methods (the *Bureau des longitudes*). But besides this, Poincaré was a best-selling author. His 1902 essay, *La Science et l'Hypothèse*, came out in a series destined for the general public and exerted considerable influence on the whole culture of the time. In Poincaré's book, there are ideas anticipating relativity that influenced Einstein, who read it in 1904, but among others also influenced was the eccentric figure of Alfred Jarry, father of Ubu's character[10].

Let us see what the meaning of the impossibility of finding an absolute criterion for space measurements is. Let us imagine being members of a community that survived a nuclear catastrophe and having to rebuild technology with our hands, having no more available the countless technical achievements we are used to: we would soon feel the need of a sample of length to use as a primary unit to measure everything else. We could cut it in wood, but after a while the sample could shrink or bend or dilate as a result of seasoning, so a wall we measured 20 metres long might look like 19

---

[8] See Bergson 1889.

[9] See Poincaré 1905.

[10] See Miller 2001, Chapter 1.

or 21 metres some time later. This could trigger practical difficulties and conflicts with the other members of the survivors' community. Then we could replace the wood sample with one of metal, but after some time we should notice that the sample marks different lengths in summer than in winter because the metal is subject to thermal dilatation. Moreover, different metals are subject to different thermal dilatation, each having its own coefficient. If two samples, one of copper and one of iron, were built in winter, the next summer the two samples would be in conflict with the measures of length, because both would be dilated by heat, but according to different coefficients, so each of the two owners would claim to be in possession of the "right" metric sample. Then, having noted this problem, we could look for a material more stable than metal, and experimentation would advise us to build a stone sample ruler: on the walls of churches, in fact, we often find sculpted samples of local measures that were in use in past times. However, we could never know that spatial measures of stone are absolutely stable: different types of stone may have a lower instability than metals (and in fact they have), but the instability of stone would be revealed by further comparisons, similar to those which informed us of the different behaviour of a ruler of copper and one of iron. Then we could choose a different fundamental sample, built not with stone but with a material that experience has shown us to be more appropriate. However, the process would not end there. When we stop in this research, what we choose as a fundamental measure is never absolutely stable: it will always be only the relatively least unstable we have found, given the complexity of the scientific knowledge we have acquired. We choose as a metric sample what seems to be the best choice for practical and theoretical purposes, without being able to know whether our choice has an absolute value. The best choice is the one that allows us to build the most coherent possible image of our experience considered as a whole. If a man chooses in winter to take a sample of ice as his ruler, it would be a crazy choice because it is obvious that in spring he could no longer solve any measurement problems, neither theoretical nor practical. Yet every choice has a gambling component which cannot be avoided.

We can always choose the length sample with rational considerations; however, the fundamental sample we have chosen through time-evolved criteria, never guarantees us an absolute value: for example, our world, the solar system and the Milky Way could be measured with margins of error which seem to us tiny, and yet with respect to the lengths measured with a sample found on some star of another galaxy they may be subject to a macroscopic phenomenon of general dilatation of their lengths for e.g. two thousand years at a time, and then to a contraction phenomenon for another two thousand years. The space lengths measured at Julius Caesar's time could be twice as long as those measured in the year 2000, from the point of view of an observer who is at a star away from us: but living in our world we would never know this phenomenon, because the fundamental metric sample we use would contract and dilate together with everything else, so we would judge stable the miles measured along the Roman consular roads and every other thing measured two thousand years ago.

So here we have the first concept of relativity, well ahead of Einstein's: we do not have any criteria for the absolute measure of space, but we always assume some fundamental sample of space that changes together with the change of scientific knowledge as a whole. So the metre of the metric decimal system, as everyone knows, was defined in the early nineteenth century as the forty millionth part of the average terrestrial meridian. But then the metre has had more than one redefinition, and today it is defined as a fraction of the distance travelled by light in a second, having measured the speed of light with a device designed according to a very rigid criterion.

1.3   Time measures: the sample clock problem

The same happens with time measurement. Before we think about it, we have the impression that the absolute measure of time is possible, but soon the consideration of concrete problems leads us to the opposite conclusion. To measure time we need a repeating phenomenon and a counter of the repetitions of this phenomenon. Thus, the fundamental clock is always based on a repetitive phenomenon (a phenomenon that has a frequency), that we take as a sample because through it the time measurements give rise to the most consistent or least inconsistent image of experience in its entirety that is possible. Also in this case we have no other criterion than the practical one.

If we imagine again ourselves as survivors of a catastrophe, we could try to use the solar day from dawn to sunset as a fundamental clock, but we would soon know it would be a bad choice: we could do it well, we could define the day from dawn to dusk as our main unit of time and then find some way to divide it into equal parts; but immediately afterwards we would realize that using the fractions of the daylight hours as time units, the time needed to cook an egg would be significantly different between summer and winter.

Then we would try to build better clocks, such as sundials and hourglasses. Apart from the obvious fact that a sundial can be used only if the sun shines in the sky, we would soon see that the times marked by a sundial and an hourglass, though both built in the best way, often disagree. In fact, due to the different speed of Earth's revolution around the Sun and the obliquity of the ecliptic, the duration of the day is not constant: every day of the year does not last twenty-four hours as the average day, but is a little longer or shorter. Differences, summing up, create a gap between true time and the time marked by the sundial that comes to more or less 15 or 16 minutes. This phenomenon is described by an equation called the equation of time. Knowing this, we would prefer the hourglass, which at this point teaches us what the character of every clock is: since we cannot do otherwise, we take as a fundamental clock some phenomenon that not only repeats itself, but in which we can neither perceive nor conceive differences of quality between a repetition and the other. It is not enough to have the repetition of a phenomenon to define a clock: it is indispensable, too, that all repetitions appear identical and indistinguishable to our judgment. Therefore, for a long time, the fundamental clock was defined on the basis of the cyclical repetition of astronomical phenomena, until in recent times it was preferred to assume a completely different repetitive phenomenon: the frequency of the microwave spectral line emitted by atoms of the metallic element caesium, which is the basis of atomic clocks. The time second was defined, up to a few decades ago, as the average solar day divided by 86,400, while today it is defined as the duration of a certain atomic phenomenon that corresponds with very high approximation, but not exactly, to the previous definition. The fundamental clock is based on a repetitive phenomenon whose repetitions appear indistinguishable without it being possible to know if they are absolutely identical in and of themselves. This phenomenon can be the Earth's rotation on itself, the Earth's revolution around the Sun, or the frequency of the caesium microwaves in an apparatus built according to a strictly defined design: there is no absolute and objective reason to prefer one choice to the other, but only the bet that we shall get a more coherent image of our entire experience.

This is exactly what is happening today: the time measured with atomic clocks was adopted as an international standard because it was considered advantageous for practical problems such as aerial navigation, but sometimes a difference is observed between the atomic time and the time measured by the rotation of the Earth, and since 1970 onward every few years, as needed, atomic time is conventionally corrected a second at a time so as to keep in sync with astronomical time. For example, a second was added to atomic time before midnight on June 30, 2012, and between 1972 and 2016 the corrections of a second were altogether 37. The correction is triggered when atomic time and astronomical time diverge by 0.9 seconds, but the problem is that nobody can predict when there will be a need for the next fix. After 2012 there was one in 2015 and another in 2016, while in 2013 and 2014 no correction was necessary. If the corrections were predictable, it would mean that we would have a better clock than both the astronomical and atomic clocks; and on the contrary, the fact that the correction is unpredictable shows us how inconceivable the idea of an absolutely correct clock is.

So here we reach the same valid consideration we reached for space: all the phenomena in our galaxy could be subject to a long cycle in which they first slow down and then accelerate with respect to some clock outside it, and we would not notice it, nor could we ever say if the "right" clock is ours or the one outside the galaxy. There is no "right" clock, but only the clock that returns the least inconsistent possible interpretation of the whole of the accumulated experience at a given time of our history. One could build a ruler with ice; in the same way, a man might observe that his canary is singing each day exactly one hundred times with an interval between one and the other tweet that seems constant to his ear, and thus he could define the division of his days based on the behaviour of the canary. The choice would be as legitimate as any other, and for a lonely man it might be the excellent one: a man who is not interested in anything outside his own world might feel at ease by choosing to use his pet's habits as a clock to decide his own actions, and cook his lunch at the fiftieth tweet, and so on. But, of course, such choice would encounter numerous practical problems in all relationships with third parties.

1.4   Simultaneity and succession of phenomena in time

When we think of two objects of any kind in relation to each other, we always represent the events relevant to the two things by placing them inside the same time. The necessity of this logical operation generates as its consequence the instinctive idea that the attribution of both simultaneity and order of succession of two events is a trivial problem. For example, let us say that a red and a white car have passed in front of our house at the same time, or that the red car has passed before the other, and we believe to "see" that the two events occurred at the same time, or which occurred before the other in a given order, with event A before event B and not event B before A. It is not so; we do not "see" simultaneity and succession, but we always decide to attribute them through a judgment that involves more elements than the simple mathematical representation of time as a line and the location of events in the same coordinate of time or in different coordinates. The mathematical definition of simultaneity and succession is very simple: given the representation of time as a Cartesian axis, simultaneous events have the same value as the coordinate t, whereas successive events have different values t_1 and t_2, which are not interchangeable. But this simple formal aspect is not enough to recognize simultaneity and succession to be the physical conditions of two events.

To become aware of the complexity of judgments about simultaneity and succession, let us first consider an example of two people, two communities or two social systems living in different worlds, knowing nothing about each other. These two systems each have their own time, which refers to the course of events in their world, and they do not have a common time. Common time arises only when the two different worlds come into a relationship, or when they are related by the thought of a third subject. In the Middle Ages it was believed that the Earth, apart from being the absolute centre of the Universe, was divided into three continents, Europe, Asia and Africa, in the centre of which there was the city of Jerusalem and around which there was only the Ocean, impossible to navigate and to cross. In this world, and in the system of geographical and temporal references of this world, in 1325 Alfonso IV of Burgundy became king of Portugal. Meanwhile in Mexico, the Aztecs lived their lives and did things in the geography of their world and in the temporal references of their chronology. It is clear that when these events happened the Portuguese and the Aztecs not only knew nothing of each other, but did not even imagine anything of each other. Someone in Portugal might have wondered, if there were men on other continents, as today we ask ourselves if there are on other planets, and the same could have been done by someone else in Mexico: but this imaginative work could lead neither the Portuguese to have a concept of Aztecs, nor Aztecs to have a concept of the Portuguese. But we know today, or we believe we know, that in the year 1325, while in Portugal the political event of Alfonso IV ascending to the throne occurred, the Aztecs founded the town of Tenochtitlán. How do we know? Obviously we get the result from comparative analysis between the European chronology and archaeological and documentary testimonies left by the Aztecs which survived the destruction of the European conquest. This, however, tells us two things: first, attributing simultaneity to two events requires that they be known by someone who can place and think them in the same reference system for time measurements. For King Alfonso IV, the foundation of Tenochtitlán is neither simultaneous, nor before nor after its ascension to the throne: it simply is nothing. Secondly, the mere representation of things that take place and change over time is not enough to attribute the character of simultaneity to two events: it is always necessary to place the events in relation to spatial facts and then correlate them reciprocally also in physical relations and not just in the geometric ones.

Let us try to remember two remote events of our early childhood: two events that have impressed us and have surprised us pleasantly or painfully and therefore are never erased from our memory, but from time to time re-emerge, making us live again the mood we experienced when they occurred. The two events are in our memory, sometimes they re-emerge solicited by other events of the present, then they return to the latent state, then they reappear. Which happened first and which later? The memory only of our mood is never capable of answering this question. If for some reason we feel the need to place the two events in exact coordinates on the line of time, we must begin by asking ourselves: what events in the outer space are associated with those two inner events? What fact generated emotional quality? The concomitant events that happened in my childhood were the cause or were the effect of my emotions? Did I make a choice because I felt some emotion, or did I feel some emotion because before I had made a certain choice? And so on: by answering these questions we can reconstruct the order of time, which the memory only of the emotional qualities is unable to return.

The objective image of the world of experience is a construction in which we need the help of distinctions in space and of the knowledge of physical relationships between events, in order to properly place distinct events in different coordinates of the time line. So, for example, while thinking of childhood memories we could say: event A must have happened before event B, because now I remember that event A happened when I lived in the first of the homes where I lived in my childhood, while event B occurred in the second home. Event C occurred after event B, because when event B was happening, Grandpa was still alive, while the memory of event C also carries the sense of strangeness I felt when I was told that Grandpa had died, and so on.

Generalizing: the time lived by consciousness is nothing more than a magma of memory contents that emerge, vanish, re-emerge, and to put order in it, we need to represent time as a spatial dimension that we could safely call the fourth dimension in a very simple sense (with little or nothing to do with Einstein). Then it is necessary to correlate the events with the spatial coordinates where they occurred, and finally it is also necessary to correlate the events among them according to physical relationships: I know something happened before another when I know the first is the cause of the second. If I do not know the physical relationship, then I can always doubt the temporal relationship.

Let us look at another example to clarify this last statement. Let us suppose we see (without audio) a movie, a comedy, in which we are alternately presented images of the country running through a window and pictures of a group of people quarrelling in a small, long but rather narrow room. The representation of the quarrel is interrupted from time to time and leaves place for short moments in which we see the country flowing out of the window, then the scene returns to the quarrelling characters. People today understand this sort of comedy scene without having to think about it, immediately telling themselves: "While a train runs through the country, some characters squabble inside the train." But to make this interpretation possible, and to pronounce that "while", we must know that there is a planet called Earth, which often we see with the aspect of the countryside, with cultivated fields and trees. That there is an invented means of transportation, said train, which travels at such a speed that the landscape seen from their interior takes a typical blurred aspect, and inside where it is possible to concentrate on different activities and to distract attention from travelling. That there are animals called men who sometimes become ridiculous by investing all their energies in futile squabbles. Without these premises we would not understand the scene, and we are able to give the judgment on simultaneity, "while," only because we are able to judge what we are seeing by knowing the fundamental physical properties that we have just stated.

We can take the opposite example and come to the same conclusion. Let's imagine seeing (always without audio) a technical documentary created to explain to technicians the operation of a kind of machinery we do not know anything about. I see a sequence in which two different viewpoints alternate. Because I know I'm watching a movie and know what a movie is, I pronounce the judgment: "The pictures I'm seeing are mounted in the film in a certain sequence." But as to the objects represented by the images, since I do not know anything about them, I do not even understand their relationships of simultaneity or succession: when the viewpoint of the camera changes, I do not know if the filmmaker intends to indicate that he or she wants to represent disjoined and unconnected events, or events occurring in the same time coordinates, or events occurring in different time coordinates with a given succession order that cannot be reversed: assigning time coordinates requires a theory of things to which the coordinates are attributed, and this theory always contains information about physical properties. If I do not know anything, then perception will only give rise to an accumulation of images in memory, which sometimes can come back if solicited by certain states of mind; but I shall never attribute simultaneity or succession unless I have a physical theory with which to interpret the accumulated images.

1.5   *Measurements of the speed of light*

Light, like all electromagnetic waves, spreads into the empty space travelling 300 million metres every second. In the Earth's atmosphere this speed is slightly lower, and in water it is significantly lower, about 2/3 of the value that is recorded in the vacuum. How can this measurement be so accurate, and what technology is needed to perform it? The answer is simpler than we would think. Already in 1677, a Danish astronomer, Ole Rømer, was able to attribute to the speed of light a value of the correct order of magnitude, something like 220,000 kilometres per second. He got this result by simply measuring for years the duration of the eclipses of one of Jupiter's satellites, Io, in the nights when Jupiter was visible from the Earth, and interpreting these observations on the basis of Kepler's laws and astronomical observations available at that time, shortly before the publication of the great Newton's treatise (1687). Rømer's reasoning was this: Jupiter has satellites, one of which is Io, orbiting around it, and we assume that the duration of the revolutions of these satellites around Jupiter is constant as well as the duration of the Lunar revolutions around the Earth. Jupiter and the Earth both perform a revolutionary movement around the Sun, but Jupiter's year is much longer than that of the Earth, and therefore the Earth and Jupiter continue to move away and converge on one another. As the Earth and Jupiter move one with respect to the other, the satellite Io enters and exits from the shadow of Jupiter: first Io is illuminated by the Sun, then it is immersed in the shadow of Jupiter, then it re-emerges and is again illuminated. Rømer observed that the duration of the eclipses of Io seen from the Earth was not constant: assuming that the duration of Io's revolution was constant, and knowing (through the laws of Kepler) the distances between the Earth and Jupiter at the time of various observations, he attributed the different duration of the eclipses of Io to the time which the light reflected by Io needed to reach the Earth. On this basis he could calculate the value we said for the speed of light, which later proved to be of the correct order of magnitude. The error of 25%, for which Rømer reached the value of 220,000 kilometres instead of 300,000, was determined by the approximate measurement of the time at night with the clocks available at that time and by other factors related to the technical resources of that era: observations were made in different places in Europe and the results were exchanged by mail. Acting more sophisticatedly, Rømer's method could have yielded a much more accurate result. For precision, it should also be said that the calculation was performed by Christiaan Huygens, with whom Rømer was in correspondence: the whole story shows the rudimentary character of the method used as a whole (see Encyclopaedia).

Other astronomical observations came to the determination of a value similar to today's accepted value, until as early as the mid-nineteenth century devices were constructed to measure the speed of a beam of light emitted by a terrestrial source and reflected by one or more mirrors to an instrument capable of measuring the duration of the travel of the beam. The necessary technology is more rudimentary than one would guess. The first and simplest of these devices was built by Hippolyte Fizeau in 1849 and yielded a value of 313,000 kilometres per second, then quickly rectified by more sophisticated devices: for the accuracy, today's accepted value is 299,792,458 metres per second and therefore the error of Fizeau's first measurement was considerable. It is worth knowing how Fizeau's first device worked to see how simple the principle was. Obviously, Fizeau knew he had to find a value of the order of magnitude of 300,000 kilometres per second, given the pre-existing astronomical determinations of this figure. So he built a large cogged wheel that had 720 teeth spaced by so many empty spaces. A projector sent a beam of light in the night to a mirror located 8.5 kilometres away, so the beam travelled a total of 17 kilometres going and coming back. The projector was placed behind the wheel teeth, so the turning wheel acted as a shutter now allowing the light to go to the mirror, now stopping it. Always behind the wheel, but on the left side of it if the projector was to the right of the wheel, an observer stared at the mirror through the interstices between the wheel teeth that passed in front of his or her eye. What happened? While the wheel did not spin, the observer saw the reflected beam in the mirror. When the wheel began to spin slowly, the observer still saw the beam. When the wheel had reached a certain speed, the observer no longer saw the beam, because it, coming back from the mirror, hit the tooth that was now standing in front of the observer's eye. By increasing the rotation speed, the beam returned visible, reaching the observer's eye through the interstice following the tooth that had passed in front of his or her eye as the beam travelled back and forth.

Since the journey time was about 1/20,000 of a second, and since the wheel had 720 teeth, to perform the experiment a small rotation speed was sufficient. It was needed that the disk could rotate at a speed of a few dozen turns per second; the first eclipse of the beam was produced at 12.6 rounds per second, so the apparatus could be constructed with the mechanical means available with the technology of the mid-nineteenth century. The calculations, however, gave a coarse value, which was corrected by about 5% by a more sophisticated apparatus soon built by Foucault.

Subsequent advanced measuring instruments determined the value of light speed more accurately, always measuring, and we shall see that this detail is important, the time of the journey of a beam that was emitted by a component of the measuring apparatus, or a source inside the apparatus, and reflected on the apparatus. That is, the instruments for measuring the speed of light in general do not measure the speed of beams emitted by external sources to them.

2. *Ether, electromagnetic theory and Lorentz's hypothesis*

2.1 *Electromagnetic Theory and law of inertia*

Once again, I warn that all notions below are simplified until being trivialized, and yet (as well as being indispensable) they are all accurate and sufficient to understand special relativity. Starting from any Encyclopaedia, you can find the necessary information to go deeper on every issue, but our exposure will make relativity comprehensible even without further in-depth knowledge.

Over the course of just a century, the theory of electrical and magnetic phenomena, whose manifestations were known since antiquity (the word "electric" comes from the Greek name for amber, the word "magnetic" comes from the name of the Greek city of Magnesia, where iron minerals were extracted) had immensely evolved, passing from Galvani's experiments with his frogs, which he mistakenly believed to be generators of electricity, to Voltaic piles and from there to electrical technology that in the first years of the twentieth century allowed the operation of electric trams and locomotives, electrical distribution networks in cities, telegraphs, telephones, Marconi's radio stations, and so on.

The theory of electrical and magnetic phenomena has been unified since, in 1820, the Danish physicist Oersted noticed, perhaps by chance, that electric current in a circuit was diverting the needles of some compasses laid on his desk. Electric charges exert forces acting in the space surrounding them, and the same does any material having magnetic properties. There are therefore, in space, electric fields and magnetic fields that interact closely: under certain conditions a current can be exploited to produce a magnetic field, under other conditions a magnetic field generates a current, and the interaction between an electric and a magnetic field is what causes the rotation of an electric motor.

The correlation between electricity and magnetism was determined quantitatively by very complex mathematical equations describing their interaction by interpreting it as a wave, with properties similar to liquid waves. During the search for these equations, in 1864, James Maxwell came to the conclusion that light is not a phenomenon of special nature, but is a part of the whole of electromagnetic phenomena, being nothing but a wave that has certain lengths: there is visible light that has wavelengths between a minimum and a maximum, there are waves of greater length than light (including those that are used for radio broadcasting) and waves of smaller length (for example, X-rays). From the study of electromagnetic phenomena, Maxwell came to determine the propagation speed of any electromagnetic wave using experimental elements and reasoning independent of those used to measure the speed of light, and found that all electromagnetic waves propagate at the same speed as light in vacuum, about 300 million metres per second. Since that time, visible light has been considered an electromagnetic wave that can have frequencies in a certain range and which generates the colours that the human eye can see.

During the development of these ideas, for which many decades were needed, a problem came to light: if electromagnetic waves are waves similar to those we observe in liquids, something must exist that moves with a wavy motion. It does not make sense to talk about sea waves without water: the sea waves are nothing more than the phenomenon of water mass movement. Something like this had to happen to allow the existence of electromagnetic waves: this consideration led to the hypothesis that wherever there is light or there are electromagnetic waves, there must also be the material medium that allows the waves to propagate, or rather, to exist. This material, hypothetical and never observed by anyone, was called the *ether*: the hypothesis dates back to the time when science was still written in Latin, and therefore the medium of light propagation took the name of *aether luminiferus*, in English *luminiferous ether*, which suggests the connotation of something archaic, immature and inadequate to the contemporary state of science. Moreover, the electromagnetic ether, "luminiferous", was a concomitant creation with recent discoveries, but the generic concept of ether was not a new hypothesis: the history of the ether hypothesis is ancient and dates back to Aristotle, who did not limit himself to the hypothesis, but was quite certain that the ether constituted the substance of the celestial world, different from the four terrestrial substances, and therefore it was the "quintessence", i.e. the "fifth substance."

But the nineteenth-century electromagnetic ether was neither a coarse concept, nor out of date; at first, the hypothesis arose only from the need to reduce the phenomenon of light to a familiar interpretation scheme, but then the concept of ether was hypothetically determined in very precise ways, according to various models, in order to attribute to it the physical characteristics that would be consistent with the properties of electromagnetic waves tested experimentally. The word chosen to give ether a name suggests a kinship with the ancient prescientific concept, but nothing more. If a modern term instead of the word of Aristotelian origin had been chosen in the 1820s or '30s, many subsequent misunderstandings would have been avoided in the divulgation of the theory of special relativity which, precisely because of the misunderstanding induced by terminology, appears to be representative of a modern paradigm of science opposed to a prescientific one.

A detail to be taken into account is that the various hypotheses attributed to the ether the ability to be everywhere, not only in the vacuum, but to share the space with each species of matter, infiltrating into the interstices between atoms. This happened because the description of electromagnetic waves suggested that they needed the same homogeneous transmission medium everywhere, unlike what happens to sound waves, which propagate in different media and cause e.g. first the air to vibrate, then the wall between two rooms, than air again, until our eardrums are reached by a sound.

The conceptual problem associated with the hypothesis of ether is that it conflicts with a conceptual stronghold of modern physics, that is, the law of inertia of Galileo and Newton. The ancients imagined, as even today many continue to imagine before receiving the first physics teachings at school, that the state of rest and that of motion can be distinguished by their intrinsic character, as solid state is distinguished from the liquid or gaseous state. Instead, Galileo and Newton taught us the principle, and laid the foundations of all mechanics on it, that rest and rectilinear motion at constant speed cannot be distinguished, if not conventionally, on the basis of a reference point chosen arbitrarily. Two bodies A and B move away from each other on a straight line and at constant speed v: we can say that A is stationary and B moves at speed v, or that B is stationary and A moves at speed $-v$, or that both move at speed $v/2$, and so on: given two bodies in rectilinear motion with respect to each other, mechanics elaborated in modern age does not possess and does not admit any criterion to decide which of them is in motion, as its absolute objective property, and which is at rest. Two people sitting in a train move both with respect to the Earth; in the meantime, with respect to the train one can be sitting quietly and the other can be moving while walking. But considering only the two persons, we say that they are simply moving the one with respect to the other, and nobody is allowed to attribute them the state of motion or rest as an absolute character. More important, nobody is allowed to say if their speed with respect to the train, or to the Earth, or to any other reference, is their "true" speed. No "true" speed, objective feature of anything, exists.

Hence follows the first principle of Newton's dynamics, the principle of inertia, for which "a body maintains its state of rest or uniform rectilinear motion, until a force does not act upon it." Because rest and rectilinear motion at a constant speed cannot be distinguished and do not differ intrinsically in anything, it is necessary to apply a force to stop or divert a body in rectilinear motion. A car switching off the engine will stop after tens or hundreds of metres from air resistance and resistance of its elastic rubber wheels, which are deformed continuously absorbing a lot of energy. A railway vehicle left without traction force will stop not after hundreds of metres but after miles, because the air resistance is less significant because it is much heavier than a car, and because the rolling resistance of the wheels is negligible, being both the wheels and rails very rigid. But a body running straight at constant speed in the empty space can travel billions of miles; it will not stop until it enters an area where a celestial body will exert a sensible gravitational influence on it.

Descartes, trying to explain this principle to one of his acquaintances, received an ironic reply: "you say that if I sit in an armchair while another moves away walking under the sun, I can say that he is at rest while I am moving?" But this witty man did not catch the point, and modern mechanics thinks differently: the condition of uniform rectilinear motion is preserved, infinitely, as long as a force does not intervene to modify it, and the condition of uniform rectilinear motion cannot be distinguished from that of rest for any intrinsic quality. Entities of any kind that are in a relative rectilinear motion with respect to each other, without anyone being able to say which is in motion and which at rest, are called "inertial systems": this expression will be used in the following.

So mechanics said, and continues to say: there is no absolute reference for the motion condition, but all references are arbitrarily chosen, given the need to study a phenomenon. If a person is in a railway carriage, I shall consider him or her moving with respect to the ground (attributing him or her a speed) or at rest with respect to the carriage (attributing him or her zero speed) as it will be better with respect to the problem I want to solve. Instead, the electromagnetic theory of the nineteenth century said (today it does not say it anymore, as we shall see below): there must be a medium of propagation of electromagnetic waves that is spread uniformly wherever our observations come, maybe anywhere in the Universe, and therefore the movement of electromagnetic waves has an absolute and unique reference system, the ether, unlike mechanical movements, who are not subject to anything like this. It is with respect to the ether that electromagnetic waves move at the speed of about 300 million metres per second, a speed always abbreviated with the letter c (from Latin *celeritas*).

The difference is not just descriptive. In equations describing electromagnetic phenomena, a constant, c, which is 300 million metres per second, represents a speed, while in the equations describing mechanical phenomena a constant speed will never appear. A man walking at the speed of 5 kilometres per hour in a train that travels at 100 kilometres per hour moves at 105 kilometres per hour on the ground: both speeds, 5 km/h and 105 km/h, are "true" because speed is calculated in a reference system chosen always for convenience and never for intrinsic reasons. Therefore mechanics cannot have a constant that represents the value of a speed. Acceleration can be a constant, and for example the acceleration of gravity on the Earth is: it is about 9.8 m/s^2 on average, and precisely 9.823 m/s^2 to the poles and 9.789 m/s^2 to the equator.

This elementary issue must be completely clear: anyone who is completely new to the subject should not continue reading without reflecting on it. Acceleration is the ratio between a speed variation and the time it takes place. If a body changes from a speed of 2 m/s to a speed of 6 m/s in a time of 2 seconds, the acceleration is equal to the difference in speed divided by the time elapsed, that is, in our case:

$$\frac{\Delta v}{\Delta t} = \frac{6\frac{m}{s} - 2\frac{m}{s}}{2s} = \frac{4\frac{m}{s}}{2s} = 2\frac{m}{s^2}$$

Acceleration is not relative, unlike speed. Suppose that the body that accelerates from 2 m/s to 6 m/s is a pedestrian walking in a train that runs at 100 m/s. From the point of view of the ground, the pedestrian passes from the speed of 102 m/s to that of 106 m/s always in 2 seconds, so we have:

$$\frac{\Delta v}{\Delta t} = \frac{106\frac{m}{s} - 102\frac{m}{s}}{2s} = \frac{4\frac{m}{s}}{2s} = 2\frac{m}{s^2}$$

Acceleration is always 2 m/s². That is, acceleration is *invariant*, while speed is related to the reference system chosen to determine it. For this reason a physical law may contain the numeric value of an acceleration and consider it a constant — for example, the numeric value of gravity acceleration — but it cannot, or rather, it should not contain the numeric value of a speed, because the "true" speed of anything can never be defined.

Thus, mechanics and electromagnetic theory had assumed a controversial principle: mechanics forbade to take as a constant any numeric value expressing a speed, while electromagnetic theory was based on equations containing the numeric value of the speed of light, but because it interpreted it as the propagation speed of electromagnetic waves in the ether, the exception to the principle of mechanics could be justified by the hypothesis that there is only one ether widespread everywhere in the Universe. With this premise, however, mechanics and electromagnetic theory could never come into a single common theory. Let us always remember that this problem, theoretically fundamental, had and has little or no practical and technical relevance: the speed of light is 300 million metres per second, while the mechanical speeds are 10, 100, 1000 metres per second: similar values related to the speed of light are, however, insignificant. So if we carry an electrical device on a carriage and to interpret its behaviour we use the laws of electromagnetic theory that contain the constant c, the fact that the carriage can move at 10, 100 or 1000 metres per second has no relevance for any technical application. This will be explained more clearly later.

2.2   *The problem of speed as physical constant*

The fact of considering any speed as a physical constant encloses a conceptual problem, and it is necessary to realize the difficulty, because it is crucial not to misunderstand the theory of special relativity.

A speed can of course be constant, but this has nothing to do with the notion of a physical constant. A physical constant is a number, a numeric value (a pure number or a certain physical quantity), that does not change in any formula. So if we assume the law of inertia and that every uniform rectilinear motion is conceived as relative to a reference system chosen arbitrarily, according to convenience, then a number representing a speed cannot appear in a physical law.

Suppose we have an electric carriage that has no speed controller: this carriage is either stationary, or moves at 2 m/s with respect to the ground. Its speed either is zero, or is 2 m/s with respect to the reference system that is meaningful to assume, that is, the ground on which the wheels of the carriage stand. If the carriage moves on the deck of a ship travelling at 10 m/s, the carriage speed will be 12 m/s with respect to the Earth, and so on. Therefore, that speed of 2 m/s, which is a property of the object, which, if powered, travels only at that speed with respect to the ground on which its wheels can roll, could never be mentioned in a physical law, because then the law would be conditioned to the arbitrary choice of a certain reference system.

Thus, the usage of the speed of light as a constant required the definition of light motion with respect to a unique reference system, which was the ether, and this is one of the reasons why Maxwell and the others who worked on electromagnetic theory in the nineteenth century considered the ether indisputable and indispensable. With this we can see that the hypothesis of the ether was not a remnant of prescientific culture, but a product of the modern scientific method, even though the hypothesis of the ether has since faded away for several reasons.

2.3   Motion with respect to the ether and the Michelson and Morley experiment

Thus, although the ether was never observed, it was thought that the ether was an inevitable hypothesis for describing electromagnetic phenomena. Now, if the ether existed everywhere and if the electromagnetic waves were moving there, then everything else had to be in motion with respect to the ether, including the Earth. At what speed? Since the Earth describes its annual revolution around the Sun at an average speed of 30 kilometres per second, its speed with respect to the ether was to be expected to be something in this order of magnitude: 30 km per second then should be added to the speed of the Sun and of the whole solar system with respect to the ether, which cannot be too high a speed, because if it were, we would observe from one year to another that we approach quickly some stars and move away quickly from others. In any case, whatever would be the overall speed, and even if the Sun were standing at rest with respect to the ether, we should expect to detect a difference of at least 30+30=60 kilometres per second in measurements separated by a six month interval, because in the second measurement the Earth would invert the sense of its movement around the Sun with respect to the sense of six months earlier.

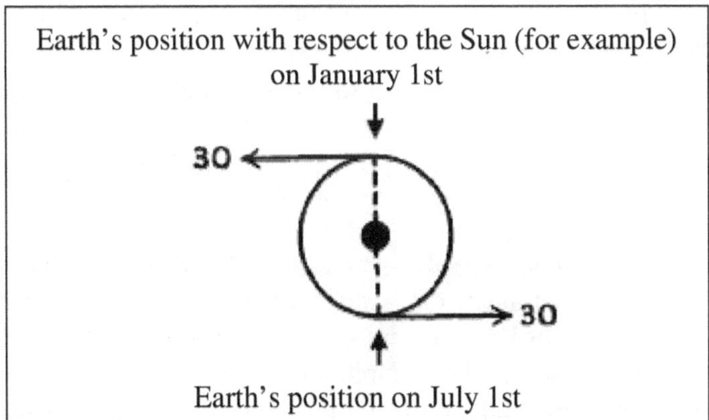

To apply electromagnetic theory in the most accurate and rigorous way, it would have been necessary to find sooner or later some method to determine the Earth's speed with respect to the ether, and then to know this data in relation to the calendar. In fact, the application of Maxwell's laws to any object moving at speed v compared to the Earth implied an approximation: to proceed with full rigour, the speed v should have been added to the Earth's speed with respect to the ether. If this data is ignored, then in principle the calculations are approximated, but given the expectation that the Earth's speed with respect to the ether is of the order of a few tens of kilometres per second compared to 300,000 for light, the approximation would be insignificant.

Since there were well-founded reasons, which we have seen, to expect a speed of the order of magnitude of at least some tens of kilometres per second, and not less than 30 km/s in certain days of the year, several experiments were designed to determine the speed of the Earth with respect to the ether. The most famous was designed by physicist Albert Abraham Michelson, who given the failure of his first attempts, associated with Edward Morley to perfect the method used. For this reason, this experiment is known as the Michelson and Morley experiment. Referring to it several times, we shall indicate it in the following with the abbreviation M+M. The experiment was repeated several times since 1881, redesigning each time the equipment needed to achieve more accurate results, and never revealing any displacement of the Earth with respect to the ether. The experiment, designed with an irreproachable logic, always yielded zero, as if the Earth were always at rest with respect to the ether, or as if the ether theory were groundless.

To understand in concrete terms the conceptual apparatus of relativity, it is not enough to know this fact. It is necessary to understand precisely the working principle of the Michelson and Morley experiment, because the method they used provides the conceptual framework of all subsequent hypotheses, including those of Einstein. One hundred years later, it is still misunderstood and a meaning that it does not have is still attributed to it. The experiment must be understood in order to understand the conclusions which are legitimate to derive from it. The simple and brief description of the experiment below is enough for our purposes: in the Encyclopaedia and in hundreds of textbooks can be found more detailed descriptions, useful to those who want to deepen their understanding of the issue.

2. Ether, electromagnetic theory and Lorentz's hypothesis

To understand the experiment of Michelson and Morley, we have first to take care of a simple introductory consideration. Let us imagine a river where water flows to our right at a given speed and imagine two motorboats leaving a mooring at point A, driven by engines that allow a certain constant speed over water.

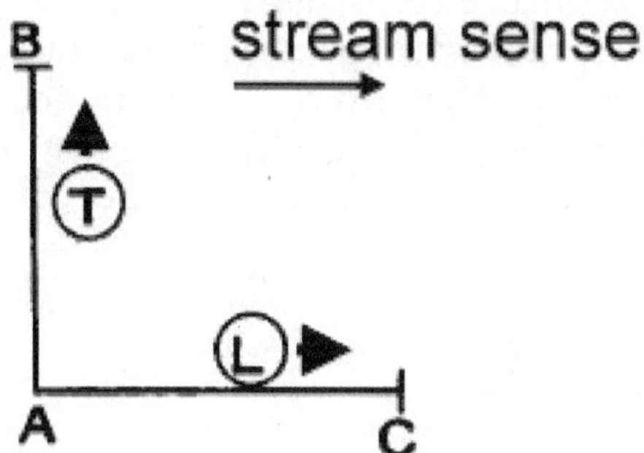

The first boat, which we call T, performs a transversal motion over the current and first goes to point B facing A on the other side of the river, then returns to A. The second boat, which we call L, performs a longitudinal movement going until point C downstream of A, then returning to A counter current.

The AB and AC distances are equal, and the boats move at the same speed. So, boat T crosses twice the current at the same speed, having the current on its side, while boat L first descends the current at a greater speed, facilitated by the water motion, then ascends the current at a lower speed, hindered by the water motion. It should be noted that if the speed of boat L were smaller than that of the current, then boat L could never return to the base at point A, but sooner or later it would reach the mouth of the river.

At first glance, one would tend to believe that the time needed to go and come back is the same for the two boats, even though boat L makes the journey forward in a very short time and that of returning in a much longer time, while boat T makes the two trips in the same time. Instead, it is easy to show that the time to go and return is always greater for boat L than boat T. Let us see a numerical example. Let's say:

- Current speed = 4 metres per second;
- Boat speeds with respect to water = 5 metres per second;

- Distance AB and AC = 9 metres.

The current goes to the right. So boat T while moving must keep the rudder steadily to the left if it wants to aim exactly at point B and reach it as soon as possible. That is, boat T is steadily in an oblique direction with respect to the current, and is constantly brought back by the current action on the line connecting A and B. So in the first second the boat moves 5 metres in the water in an oblique direction with respect to the shore being pushed by its engine, and 4 metres downstream being transported to the right by the current. The result by the Pythagorean theorem is that it moves at 3 metres per second in the AB direction.

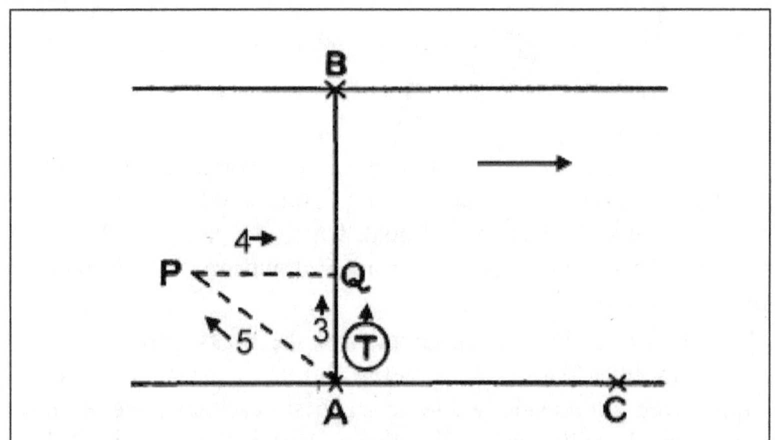

Boat T in the first second points towards P, and is dragged by the current towards Q. So it travels AQ = 3 metres.

In fact, in the first second of the journey, the thrust of the engine and the current give a composed effect, and the result is that the boat moves towards B along the AQ segment along x metres, and we have by Pythagoras:

$$AQ^2 + QP^2 = AP^2; \qquad \text{or: } AQ^2 + 4^2 = 5^2;$$

hence

$$AQ^2 = 5^2 - 4^2; AQ^2 = 25 - 16; AQ^2 = 9; AQ = 3.$$

So since boat T runs 3 metres every second, to reach point B on the other bank, which is 9 metres far, it takes 3 seconds; then given the symmetry of the situation, to return it takes another 3 seconds, so all it takes is 3+3=6 seconds.

Instead, boat L first descends the current, so its speed is higher than that of the current, and it travels at 5+4=9 metres per second. To go from A to C, running 9 metres, it takes only one second. On the return, the speed of the current is subtracted to the speed of the boat, so it travels at 5-4=1 metre per second, and so to return from C to A it takes 9 seconds. Throughout the longitudinal boat takes 1+9=10 seconds, which is more than the 6 seconds necessary for the transversal boat. The relationship applies to any value as long as the speed of the boats is higher than the current speed. Otherwise things would be completely different, because the boats could not return to point A.

The Michelson and Morley experiment was designed taking into account this relationship between the longitudinal and the transverse motion. In it, a beam of light is decomposed by means of a prism into two beams which move each one for a few metres in two directions that form a right angle. Then the two beams are reflected by mirrors and return to a common point. If the beams move like waves of the ether, and if the Earth, or the equipment, moves with respect to the ether, we have the same situation described in the example of the two boats: the Earth's speed with respect to the ether is the speed of the current, and the speed of the two beams is the speed of the two boats with respect to the water. We expect, of course, that the speed of the current, i.e. of the Earth relative to the ether, would be much smaller than the speed of light, that is, the speed of the two beams. The two beams, after travelling in rectangular directions and appropriately reflected, converge to a point where they should arrive at a different speed with respect to the equipment, so with different frequency and wavelengths for each of the two; this difference should produce a known and easily observed phenomenon, the interference fringes. The phenomenon of interference is this: two beams of different wavelength and frequency converging at one point produce monochromatic or coloured shapes similar to the shape of the rainbow, that the human eye distinguishes easily also for very small wavelength and frequency differences. The visual representation has this aspect:

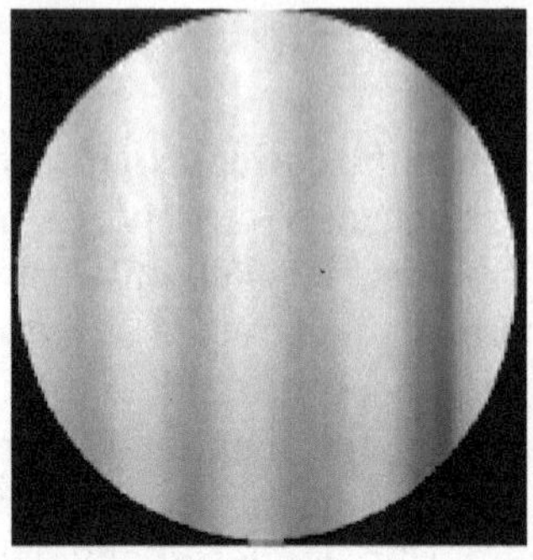

There is a complex theory of interference (see Encyclopaedia), but it is a theory that was consolidated and easy to verify already in the nineteenth century, because in order to produce interference it is enough to converge beams of different colours. To understand the principle of the experiment, it is enough to know that if the Earth moved with respect to the ether at a speed, for example, of the order of magnitude of the revolution around the Sun (tens of kilometres per second), or even much less, the theory ensures that the interference fringes would be perceivable with evidence. The experiment would give a definite result, and the study of the fringes would make it possible to calculate the Earth's speed with respect to the ether at different times (remember that it should surely change during the year).

The apparatus (of which you can find accurate descriptions in the Encyclopaedia and also a single old surviving photograph) worked as follows: the complex formed by the source of the beam of light, the prism that split it into two orthogonal beams, the mirrors that sent the two components back on a screen, and finally the screen for observation of the interference fringes, was mounted on a circular support that would allow it to rotate. The whole apparatus floated in a mercury tank, which would ensure that vibrations from the outside environment were best absorbed. At any time, when the complex of the apparatus was in any position, switching on the lamp would have made some interference fringes visible. The Earth would move in the ether by generating a current exactly equivalent to a river stream, called the "wind of ether". Turning the apparatus, one of the two beams sooner or later would move longitudinally with respect to the Earth's motion in the ether and the other transversally, and then rotating by another 90 degrees the situation would be reversed. Thus during the rotation of the apparatus the interference fringes would change continuously their aspect, and they should reveal the moment when the two beams were one in the longitudinal position and one in the transverse position with respect to the motion of the Earth in the ether.

The simplest possible schematic representation of the device is that of this figure (if necessary, there are many other useful illustrations on the Internet). To understand the operation of the equipment, consider that the mirrors are inclined in the appropriate way so that the beams do not interfere, and that only in the last part of their travel do the beams converge and create interference fringes:

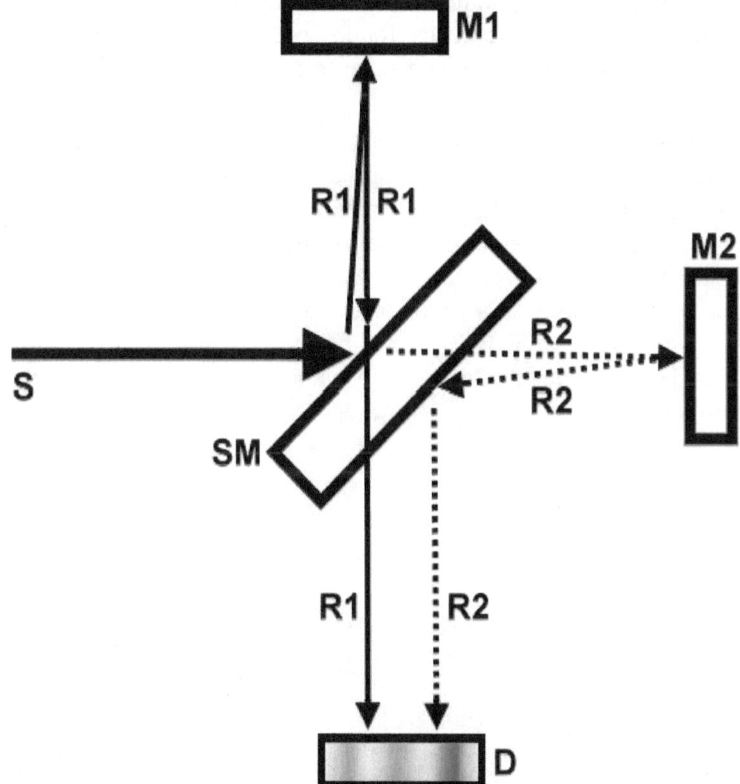

A beam is emitted by the source S and is directed to the semi-reflective SM mirror. Part of the light is reflected by SM towards the mirror M1, while another part crosses SM and continues towards the mirror M2, so the beam is split into two beams R1 and R2. The beam R1 (continuous line) is reflected by SM towards M1, there it is reflected again, it crosses SM and reaches the screen D. The beam R2 (dashed line) crosses SM, reaches the mirror M2, is reflected back, reaches SM and is reflected towards D. R1 and R2 converge towards D by mixing together, so they interfere with each other and generate interference fringes on the D screen.

The result of the experiment was surprising, given the confidence that was granted to the existence of the ether. The interference fringes, which were seen because the inevitable approximations in the construction of the apparatus implied that they were generated in each case, rotating the apparatus never changed their aspect. For the sake of precision, the differences found in fringes were equal to about 1% of the expected ones and were to be fully attributed to the approximation in the construction[11].

Then it was assumed that a chance had liked the Earth to be at rest in the ether at the first attempt: at a given moment of the year this could happen, but given the Earth's movement around the Sun, waiting some days or months a change of aspect of the interference fringes should manifest itself. But it never occurred. Discarding the interpretation that this was a proof of the ancient geocentric theory (hypothesis that for completeness must be mentioned, but obviously no one takes into consideration), it became necessary to find an explanation of the phenomenon.

---

[11] See table in Resnick 1958 p. 24.

2.4   Propagation of electromagnetic waves

Before discussing the negative result of the M+M experiment, a general premise is needed. We can represent two ways of propagating light: like a wave in the ether, or like a movement in the vacuum (or in the air or in transparent substances) of something that moves away from the source that emits it.

In the first model, light and electromagnetic waves behave like sound, which is caused by vibrations of air or water where sound propagates, while in the second model they behave like a bullet thrown by a weapon. In the first model, electromagnetic waves move with respect to the reference system given by the ether, while in the second model they move in a reference system that originates in the emitting body. According to this use, the two models will be called "ether theory" and "emission theory" respectively. Emission theory is sometimes also referred to as "ballistic hypothesis", because electromagnetic waves would behave as bullets.

In the ether theory, the speed of the wave source must not be added to the speed of the waves. It is very easy to understand that this is so: suppose the Sun emits a light beam towards the Earth beneath it, and in the meantime it moves with respect to the ether at any speed. For simplicity, suppose the Earth to be at rest in the ether. The beam will reach the Earth in the eight minutes necessary regardless of the Sun's motion during that time.

Situation at the moment of the emission of the beam.	Situation 8 minutes later.
The beam is delayed by the Sun to the ether, in whose frame the Sun is in motion.	The beam reaches the Earth after having travelled at speed c with respect to point A, the Sun has moved in the ether and is elsewhere.

The light beam moves at speed c with respect to the ether, while the Sun may be at rest or move in it at any speed and in any direction: in any case the beam moves at speed c with respect to point A in the ether where the Sun was at the time of the emission, and in the meantime the Sun may be stationary at point A or moving with respect to it in any direction without changing the speed of the beam.

The situation would be the same if both the Sun and the Earth were in motion in the ether. Some of the beams emitted by the Sun would travel a trajectory so after eight minutes they would reach the Earth's surface, and the speed of the Sun, that is, of the emitting body, would be irrelevant:

Situation at the moment of the emission of the beam.	Situation 8 minutes later.
The beam is delayed to the ether, in whose frame the Sun and the Earth are both is in motion.	The beam reaches the Earth, the Sun has moved in the ether and is elsewhere, the Earth did the same.

So in the first model the speed of the source of electromagnetic waves is always irrelevant. Light and electromagnetic waves propagate in the ether at speed c independently of the speed of the source. This situation is hypothetical for electromagnetic waves (in fact, this is the old hypothesis), while it is certain and unquestioned for the sound: sound propagates in the air at 340 m/s, and this speed is not to be added to the possible speed of the source. If a locomotive whistles, the sound moves towards us at its usual speed with respect to the air, not at the overall speed resulting by adding its speed with that of the locomotive.

In the second model, that of the emission theory, the ether is suppressed and it is assumed that the light moves in the vacuum. In this case, a beam of light would move with respect to the body that emits it by inheriting its inertia and its relative speed. In the case of a bullet this is exactly what happens. Let us suppose a gun gives the bullets an acceleration so they exit the barrel at the initial speed of 900 m/s, then they slow down under the effect of friction with air and gravity. This speed is added to that of the gun. If a man on the ground fires a shot, the bullet exits the barrel at a speed of 900 m/s with respect both to the gun and the ground. If, however, a gun is fired within a train running at 100 m/s in the direction of the train's motion, then the bullet speed is 900 m/s with respect to the train, but 1000 m/s to the ground. But if, for example, the train goes south and the bullet is fired to the east, then the bullet exits the gun at a speed of 900m/s with respect to the axis of the barrel to the east and 100m/s with respect to the direction of the train to the south, perpendicular to the axis of the barrel. The situation is this:

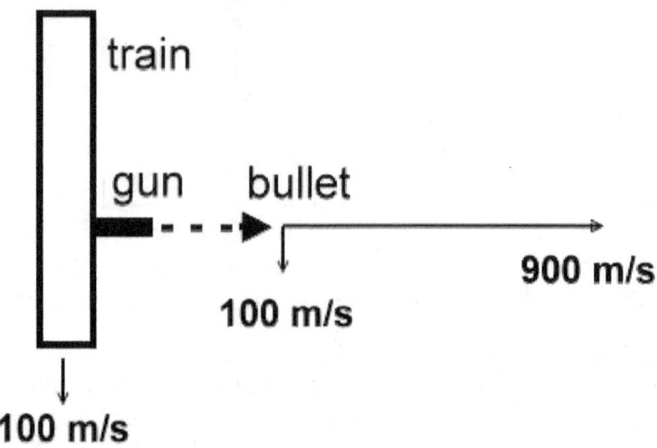

Given this initial situation, if there were no air resistance and no gravity, and if both the bullet and the train continued their rectilinear path at constant speed, the bullet would always be aligned with the gun barrel because it would proceed endlessly to the east at 900 m/s and to the south at 100 m/s together with the train. In the physical reality of the terrestrial world, however, the bullet would offer the edge to the air resistance in the south direction and would thus lose more speed in the south direction than east direction.

As for light and electromagnetic waves, in the frame of emission theory the same thing would happen: a light beam would be aligned with the emitting body until it is diverted by any cause, or until the emitting body changes the speed at which it was moving in a straight line at the time of the emission of the beam. The Sun emits beams in every direction, and a small part of them is destined to reach the Earth and there to stop converting to other forms of energy. How would a beam of light destined to reach the Earth behave in the hypothesis of the emission theory? Since the Earth's motion with respect to the Sun is considered relative by the law of inertia, we can describe the situation in infinite different ways. We can consider the Sun at rest and the Earth in motion, and then the travel of the beam of light would be interpreted and described as follows:

Situation at the moment of the emission of the beam.	Situation 8 minutes later.
The beam starts from the Sun considered at rest.	The beam reaches the Earth after having travelled at speed c downwards. The Earth has moved to the left and has met and captured the beam.

In this first case we shall say that the beam has travelled at speed c with respect to the Sun, and the Earth moved at speed v to the left, intercepting the beam after the necessary time.

But with the same right we can consider the Earth at rest and the Sun in motion. In this case the situation is described as follows:

Situation at the moment of the emission of the beam.	Situation 8 minutes later.
The beam starts from the Sun that is moving to the right. The Earth is considered at rest.	The beam started in point A reaches the Earth after having travelled at speed c downwards and v rightward.

In this case we shall say that the beam travelled at speed c with respect to the Sun and at the speed resulting from the perpendicular components c and v with respect to the Earth, describing a diagonal trajectory along the line connecting point A and the Earth. The cases are infinite (not just the two ones of the example): the Sun and the Earth can also be considered both in motion at any greater speed than zero and less than v, provided the sum of the two speeds is v.

But the Sun (whose motion is elliptical, but here given the brevity of its path in eight minutes can be assumed to be rectilinear) in any case is at the origin of the beam when the beam reaches the Earth, unless the beam is diverted by forces of any kind and nature during the journey, or the Sun abandons its rectilinear path or changes its speed. As long as the beam moves with the inertia received at the time of emission, in the emission theory we cannot conceive that the Sun abandoned point A at which it was when it emitted the beam, because in order to conceive this we would need an absolute reference system, different than the Sun, to determine the motion of the beam. Hence we can never form this representation:

Situation at the moment of the emission of the beam.	Situation 8 minutes later.
	The beam reaches the Earth, the Sun has moved with respect to the path of the beam: NO!

If we admitted this representation, we would be postulating a reference system that allows to determine a displacement of the Sun from the point where it was when it issued the beam, so we would be out of the theory of emission, returning to the hypothesis of ether, or maybe postulating some other absolute spatial reference.

For all this, in the model of emission theory, the speed of light is the sum of c and of the relative speed of the source, whereas in the theory of ether the speed of the light is always and only c with respect to the ether, and we should became able to determine the speed of the celestial bodies with respect to the ether. When reading the comments on the M+M experiment in the following pages, we shall need to consider this distinction and some other factors. First of all, that at the time of the experiment, the only scenario taken in consideration was the theory of ether. Emission theory was not taken into account for many reasons that we shall see. And then, it must be considered that in addition to the null result of the M+M experiment, it occurred that every device designed to measure the speed of light had always returned the same value, and this should not have happened because the measured speed should have varied in relation to the Earth's speed with respect to the ether. But precisely because light speed measurements always gave the same result, the device of the M+M experiment was designed not to measure the speed of light, but to accurately detect only the Earth's motion with respect to the ether.

We can conceive the project of constructing equipment to measure the speed of a beam of light emitted by the equipment itself, or to measure the speed of a beam emitted by an external source, for example the Sun, with respect to which we know with certainty that we are always in motion at a given speed. The second case, however, is technically far more complex than the first one since to measure the speed of light, a coherent beam having a precise wavelength must be generated, and therefore measuring a beam generated within the apparatus gives much more accurate results compared to those that would be obtained by measuring the speed of beams coming from the Sun or from another star.

However, it is necessary to understand that these two types of measure conceptually would have different consequences in the theory of ether and in that of emission.

In the theory of ether, any beam is considered in motion with respect to the ether, and therefore the choice to measure a beam emitted by the measuring apparatus or by an external source in motion is irrelevant. The beam source delivers the beam to the ether, and then it can stay at rest at that point of the ether or it can move towards somewhere else, without affecting the speed of the beam. The expectation is that the measurements of the speed of light reveal the speed of both the Earth and the observer not with respect to the source, but with respect to the ether, and this does not happen, and therefore the problem of ether was born.

In emission theory, the beam is considered in motion with respect to the emitting body and therefore the difference between the measurement of a beam emitted by the measurement apparatus itself or by an external source in motion is relevant. If the source of the beam is at rest with respect to the measuring apparatus, the expectation is that the measurements always return the same value, consistent with the central hypothesis of the theory, and that is exactly what happens in the experiments. If the source of the beam is outside of the apparatus and in motion with respect to it, the expectation would be to detect the speed of the observer or of the Earth with respect to the emitting body: but this is still indeterminate, because it does not appear that experiments have been carried out up to date with the evolution of technology to determine the speed of light emitted by external sources in motion, like the Sun.

But then if the emission theory had been taken into account, the M+M experiment would not matter, because it is based on light beams emitted by the equipment and reflected within it, and therefore given the fundamental hypothesis of the emission theory, it would not have to detect any interference between orthogonal beams; for the same reason any device based on beams generated inside it should always return the same value for the speed of light. Thus, the M+M experiment is significant and constitutes a problem only when the ether is assumed, while the null result would be expected and it would not be a problem if the hypothesis of the emission were adopted. More specifically, the M+M experiment is not enough to confirm the emission theory, but it is not a proof that falsifies the emission theory: it puts only the hypothesis of ether into difficulty.

The following table summarizes what we saw in this section. The results found in the experiments obviously do not depend on the hypothesis that is adopted and are unique for both hypotheses. While the measurement of beams emitted within the equipment always returns c, it is not known today whether the measurement of beams emitted by a source in motion with respect to the equipment returns different results, and therefore overall we have this situation:

	Beam emitted by the equipment	*Beam emitted by a moving external source*
Result found in experiments	c	None. No systematic check has been carried out.
Expected result according to ether theory	c + speed of the equipment with respect to the ether. Disagrees with experimental results.	c + speed of the equipment with respect to the ether.
Expected result according to emission theory	c. Agrees with experimental results.	c + speed of the equipment with respect to the source of the beam.

While in the hypothesis of the ether we have the anomaly for which the result of the measurements is always c, in the case of the emission theory this is not an anomaly, but it is just the expected result, at least until the experiments are made with beams emitted by sources at rest with respect to the apparatus.

2.5   *Lorentz's "physical" relativity theory*

Returning to the M+M experiment and to its negative result, the simplest thing that comes to mind is to say: but then there is no ether and light travels simply in vacuum, according to the ballistic hypothesis that is consistent with the negative outcome of experiments and measurements. Instead, physicists sought to maintain the concept of ether, already hypothetical, with additional hypotheses. There was, at that time, still a cultural resistance to the idea that a motion could take place in the vacuum ("natura abhorret vacuum"), but the reason why the new view was not easy to adopt, was by no means simple reluctance to discard the ancient and prescientific prejudice of the fear of the idea of empty space, the so-called *horror vacui*. This is always suggested by the more or less popular expositions of Einstein's theory, which all say that the assertion of the ether was a traditional and antiquated prescientific remnant, while its denial would be an up-to-date and modern attitude, so explicitly suggesting a heavy value judgment in favour of the denial of the ether. It is not right to consent to this value judgment, which is unfair because the electromagnetic theory of the late nineteenth century was a mature science, and is misleading because it loses sight of the serious problem that relativistic theories, not just Einstein's, tried to solve.

The important reason why it was difficult to abandon the hypothesis of the ether was that this would require a revision of the form of the electromagnetic theory equations, in which, we recall, the speed of light appears as a physical constant. If we remove the ether which is the preferred frame for the speed of light, by what right would the formulas contain the expression c? The speed of light will become relative to the chosen frame, as the speed of any kind of thing from the mechanical point of view. It can also be argued that the speed of light in vacuum is always c with respect to the body that emits it; but if the body that emits it moves at speed v with respect to another frame, the speed of light emitted will become $c+v$ in the new system, and formulas of general value will not be allowed to contain the expression c, if not at the cost of becoming approximate and theoretically incorrect. Let us focus on this point: if a physical law contains the expression c, which represents a speed, and if we cannot say with respect to what constant frame that speed is calculated, the law becomes an expression of a poor and illogical theory. Since we are dealing with the tremendous speed of light, the law can remain valid with good approximation as the speeds of the mechanical and astronomical world are all far smaller than c: and that is exactly what happens. But the theory is weakened by the fact that it contains a conceptual indetermination in its foundations.

Imagine being inside a train travelling 100 metres per second and reading with the light of a lamp above our head, which emits light in different directions. If the light is transmitted by the ether, its absolute speed is c, and c can be regarded as a physical constant experimentally determined as others. The speed of light is always c with respect to the ether, while with regard to the Earth or to a train that moves on the Earth it should be corrected by adding or subtracting the speed of the Earth or of the complex of the Earth and the train with respect to the ether. Instead, if the light moves at speed c with respect to the lamp that emits it inside the train carriage, then practically the situation is similar, but electromagnetic theory no longer has the right to use c as a physical constant. If the train moves at speed v, then the speed of the light emitted by the lamp is equal to c inside the train, but it is equal to the $c-v$ with respect to each point on the Earth from which the train is moving away, and to $c+v$ with respect to every point the train is approaching. The value of c becomes the expression of a physical law which asserts that light always departs from its source at that speed, but is no longer a constant that holds the same value in any calculation and any situation.

Then, how could a physical law of universal value mention the constant speed c, which is 300 million metres per second, knowing full well that this speed is sometimes 300 million plus 100 metres per second, or else 300 million less 100, and so on, and being aware that none of these speeds has the right to be privileged, and that none of these speeds is the "true" one? We can comfort ourselves with the fact that from the point of view of technical use, little or nothing changes: mechanical speeds on the Earth, in the solar system, and elsewhere are always irrelevant if compared to the speed of light. But we do not have a good theory.

2. Ether, electromagnetic theory and Lorentz's hypothesis

The difference between the conceptual framework of the ether and that of the emission is thus: in the case of ether c is the "real" speed and the theory is exact and consistent, although it is approximate in the application because we do not know the speed of anything with respect to the ether; instead, in the case of the emission theory no speed is ever "true", so not even c is, and therefore the theory makes an arbitrary choice against its premises and is inconsistent, even if the approximation in the application is the same as the first case and is devoid of practical consequences.

Therefore, the reason why scientists tried to preserve the concept of ether were not to indulge in an old-fashioned cultural habit, but was not to put into question the electromagnetic theory elaborated throughout the nineteenth century and enabling now science to master electromagnetic phenomena. From a theoretical point of view, the problem was serious. From a practical point of view, it was not relevant to mechanical speeds that are very small compared to that of electromagnetic waves, but it was relevant in particle experiments, whose earlier ones had already been performed and in which speeds near to that of light are dealt with.

In order to preserve the concept of the ether, some physicists elaborated a hypothesis that as admitted by the same creators was completely *ad hoc* and was not supported by any experimental data, but which was entirely plausible from a logical point of view. The hypothesis was that everything that moves with respect to the ether is subject to a contraction of its spatial lengths and a proportional slowdown of all its movement and change processes. This phenomenon is known as FitzGerald's contraction, or Lorentz's contraction. The physicists who worked on this hypothesis were first George Francis FitzGerald, a relatively obscure figure, and then the well-known Hendrik Lorentz and Henri Poincaré, as well as Joseph Larmor, who completed the theory of the so-called "Lorentz transformations", about which we shall discuss soon.

The hypothesis consists of imagining that celestial bodies, the Earth, and everything else are deformed, complying with a precise factor, when moving with respect to the ether, and then resume the primitive measures when they return to rest. Lorentz elaborated an electron theory conceived so as to make plausible the phenomenon of this deformation at the atomic level.

Of course, the numeric factor of this contraction is irrelevant and virtually impossible to detect (calculations made according to the hypothesis show that it is smaller than it would be expected at first glance: we shall see it in the next pages). But one must imagine that the hypothesis would be conceivable even if the speed of light were much lower than it is and therefore the phenomenon were more relevant. A sphere moves with respect to the ether: in the direction of motion the sphere is contracted and becomes an ellipsoid. No one perceives it, because all things undergo the same contraction. I walk by moving with respect to the ether: because of the fact that I move I am deformed in the size of the motion and meanwhile all my internal, mental and physiological processes slow down in the same proportion: however, I do not notice anything because everything is subject to this deformation, in which all the spatial lengths in the direction of motion in the ether together with the duration of all phenomena are contracted. Clocks slow down, biological clocks slow down, and I notice nothing, not because the contraction factor is tiny, but because all things contract with the same factor, including the tools to measure them, so the measurements of all things remain the same. The same thing would happen if the phenomena of the whole world visible to us slowed down or accelerated all together with respect to the frequency of a clock outside our world: we would not know anything about it.

As a first hypothesis, only the contraction of the lengths was assumed, because given the way the apparatus of Michelson and Morley operates this was sufficient to explain the negative outcome of that experiment. In fact, the experiment of Michelson and Morley does not include any clock nor any time determination, but is based on spatial measurements: the beam split in two orthogonal beams should travel different lengths due to the Earth's movement with respect to the ether and therefore the two beams recombining should give rise to visible interference fringes. The time taken by the two beams is the same, while the travelled path is different and the device does not need to perform time measurements with clocks, because the phenomenon of the fringes formation should occur for the whole duration of the flow of both beams. If the arm of the apparatus that moves with respect to the wind of ether is contracted, the motion of the whole apparatus with respect to the ether is compensated and the null result of the experiment is explained.

From the spatial point of view, the phenomenon is very simple:

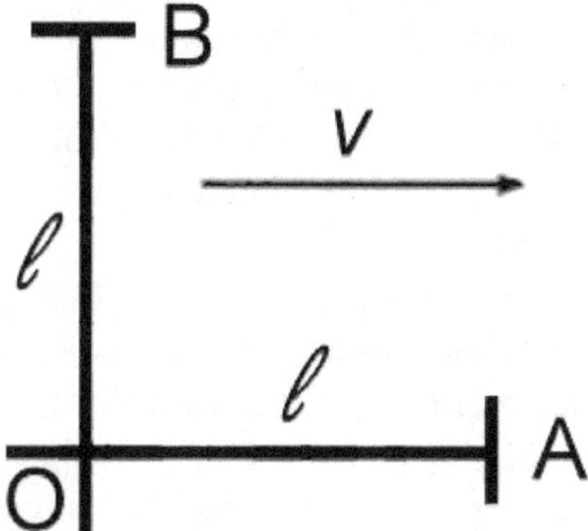

Suppose that OA and OB are the arms of the M+M apparatus, both being long l, that O is the prism that splits the light beam in two and that A and B are the mirrors. All the apparatus moves in the ether to the right at speed v. The light beam ranging from O to B travels the distance l not altered by the movement in the ether. The beam ranging from O to A should travel a greater distance than l, because the mirror A moves to the right during its journey. But the OA arm is contracted, and its length becomes less than l, exactly what is needed in order that the light travels the length l overall. A very important consequence must be observed immediately: in this framework not only is the M+M experiment explained, but also explained is why light speed measurements by means of apparatus like Fizeau's always give the same value c, without revealing differences depending on the motion of the Earth in the ether.

But Larmor immediately added the hypothesis of the contraction of the durations of phenomena over time, which was suggested by the development of a more detailed hypothesis that both Lorentz and himself were working out to explain the phenomenon of contraction at the atomic level. Lorentz and Larmor did not limit themselves to state the length contraction hypothesis, but developed an articulated theory of electron behaviour, designed to explain the phenomenon of length dilatation, and Larmor found that for inner coherence reasons this theory of the electron should include the contraction of the durations. These atomic theories were abandoned a few years later; what matters is that Lorentz's theory, if desired, is also legitimate to predict the duration contraction, although from a logical point of view this additional hypothesis is not necessary. According to the theory, everything contracts if it is in motion with respect to the ether, and nobody realizes that all phenomena are contracted in space by the same factor. Thus, it is possible to assume that even the durations of phenomena in time are contracted: no one would notice even the temporal contraction.

Hence assuming that both lengths and durations are contracted, those who undergo the double contraction of space and time will not notice anything. Lorentz expresses this as follows[12]:

> It is of importance not to forget that, in doing all that has been said, the observer would remain entirely unconscious of his system moving (with himself) through the ether, and of the errors of his rod and his clocks.

The leading actors of this hypothesis were not enthusiastic: they all maintained strong awareness that this was an *ad hoc* hypothesis, adopted with little conviction, awaiting a theory better connected to experience. However, this was a hypothesis to which no fundamental incoherence could be objected to, and which already included the central idea of relativity, that space and time can dilate or contract. If all phenomena that take place over time slow down by the same factor (i.e., in the same way and by the same degree), we can say reciprocally that their time has dilated, and the same for space. So now we have sufficient elements to understand in what sense Lorentz and Poincaré used the concept of *local time* before Einstein: the concept is hypothetical, but the hypothesis of spatial and temporal contraction entails the hypothesis that phenomena can slow down and accelerate periodically, under certain conditions, and therefore time measures can become relative.

Let us conclude with a leap to a theme that we shall have to discuss in detail talking about Einstein's theory. Everyone knows that the canonical and extreme example of the theory of special relativity is the so-called twin paradox: a traveller flying in space at a speed close to that of light and then returning to the base would come back

---

[12] Lorentz 1916, p. 226.

younger than those who stayed on the Earth. The example expressed in this form is fantastic, but conceptually it is consistent with Lorentz's hypothesis, and it is perfectly understandable if we add the necessary specification: if there were a floating launching base anchored to the ether, and if a traveller moved in the ether and then returned to the starting point by the same road in reverse, at the time of his return he would find his clock back with respect to those inhabiting the launch base, who were stationary with respect to the ether. And since the slowdown would have affected phenomena of every kind inside his space ship, both mechanical and biological systems, if the traveller had a twin on the return he would find himself younger than his brother of a more or less relevant amount of years.

2.6   *Lorentz contraction factor*

We need little and simple math to understand Lorentz's hypothesis, and this little mathematics is exactly the same that will be used by Einstein, who will however give a different interpretation of the problem and will attain a specific theory. But Lorentz's formulas will be used and applied by Einstein without changing anything of their form. Therefore, the work done to assimilate the mathematical consequences of Lorentz's hypothesis will not be repeated when the same formulas will reappear in Einstein's work.

We know that according to the hypothesis the Earth and everything else in motion with respect to the ether undergoes a physical contraction in the longitudinal direction to the motion, and no contraction in the transverse direction. Quantitatively, what values has this contraction? Getting the formula is within reach, and the procedure is described in all the special relativity treatises. The reasoning is similar to what we have outlined in relation to the problem of the two boats that make an equal journey on a river, leaving a common starting point and returning, moving one in the transverse and the other in the longitudinal direction with respect to the current. The second of the two boats takes a longer time than the first to go and come back along a path of the same length. Let us imagine the same for two beams of light starting from a common point and moving to a mirror that reflects them, going one in the transverse direction with respect to the ether wind, and the other in the longitudinal direction. The second beam should take up more time than the first; but since the experiment of Michelson and Morley has pointed out that this increased duration is never detected, we hypothesize that all things undergo a contraction in their longitudinal dimensions with respect to the movement in the ether.

Rigorously, we know that the Michelson and Morley experiment does not include time measurements; however, the experiment provides information about the speed of the beams, because the two beams do not cause interference after travelling along orthogonal directions, and thus behave as if they had travelled at the same speed: moving with respect to the ether did not change the speed of any of the two beams, as was expected to happen.

2. Ether, electromagnetic theory and Lorentz's hypothesis

Let us imagine, then, that from point O a beam starts to point A, and then this beam is reflected in a mirror and returns to O. Another beam starts from O to go to B, and there it is reflected and returns to O. The OA and OB segments have the same length l.

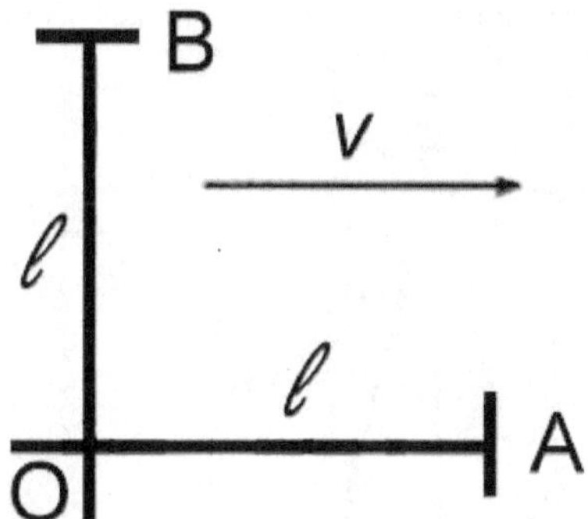

The whole system moves to the right with respect to the ether, in the OA direction, at the speed v.

Let us see how the beam behaves in the direction of OA, longitudinal. On the way forward, the OA journey takes place at speed c-v, because the journey takes place in the ether at speed c, and the mirror in A moves away from the beam at speed v. On the way back the travel AO takes place at the speed c+v because the source point O moves towards the beam, always at speed v. So the time to go and come back through OAO is:

$$\frac{l}{c-v}+\frac{l}{c+v}=\frac{2lc}{c^2-v^2}\ (1)$$

The average speed is (obviously):

$$\frac{c+v+c-v}{2}=\frac{2c}{2}=c\ (2)$$

The total travelled space is not $2l$: this could be thought at first glance, but we have to take into account that in the first part of the trip the mirror moves away from the beam, while in the second part the source point O moves towards the beam, and then the total path space OAO is, multiplying (1) for the mean speed:

$$OAO=\frac{2lc\,c}{c^2-v^2}=\frac{2lc^2}{c^2-v^2}=\frac{2l}{\dfrac{c^2}{c^2}-\dfrac{v^2}{c^2}};$$

hence:

$$OAO = \frac{2l}{1 - \frac{v^2}{c^2}}; \ (3)$$

Instead, the path of the beam OBO happens as shown:

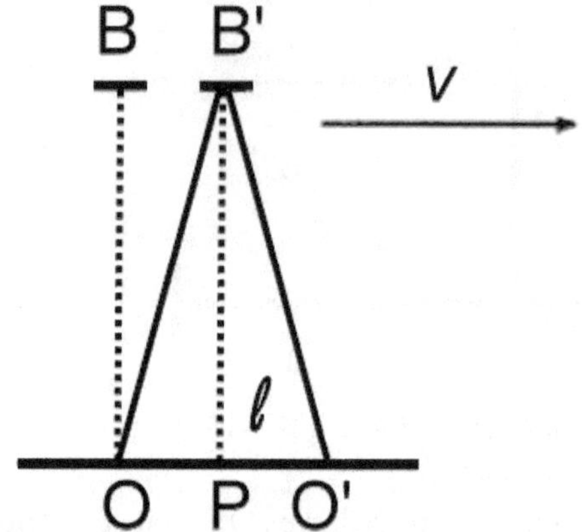

The whole system is transported to the right in the ether, so the beam travels first along the distance OB', and after reflection along the distance B'O'.

The overall time is:

$$t = \frac{OB'O'}{c} = \frac{OO'}{v} \ (4)$$

because while the light beam travels the segments OB' and B'O' at the speed c, the system travels through OO' at speed v. Since speed v is supposed to be constant, the triangle is isosceles and we can divide by two getting:

$$\frac{OB'}{c} = \frac{OP}{v} \ (5)$$

By Pythagoras we have:

$$OB'^2 = l^2 + OP^2$$

hence:

$$OP = \sqrt{OB'^2 - l^2} \ (6)$$

Now we want to find OB'. Replacing in (5) we have:

$$\frac{OB'}{c} = \frac{\sqrt{OB'^2 - l^2}}{v}$$

from which through simple passages is derived:

$$OB' = \frac{cl}{\sqrt{c^2 - v^2}};$$

The total space travelled by the transverse beam is twice OB' (the triangle is isosceles), so the overall distance OB'O' is:

$$OB'O' = \frac{2cl}{\sqrt{c^2 - v^2}} = \frac{2l}{\sqrt{1 - \dfrac{v^2}{c^2}}}; (7)$$

At this point, considering (3) and (7), we see that the ratio between longitudinal and transverse displacement is:

$$\frac{OAO}{OB'O'} = \frac{\dfrac{2l}{1 - \dfrac{v^2}{c^2}}}{\dfrac{2l}{\sqrt{1 - \dfrac{v^2}{c^2}}}} = \frac{1}{\sqrt{1 - \dfrac{v^2}{c^2}}}; (8)$$

and this result is the Lorentz contraction factor, which we shall always represent with the letter γ, gamma:

$$\gamma = \frac{1}{\sqrt{1 - \dfrac{v^2}{c^2}}} (9)$$

and which serves to convert the measurements in the longitudinal direction of the motion with respect to the ether to the measures that would be taken in the transverse direction. According to the contraction hypothesis, moving in the direction of the ether all things undergo a contraction of factor $\sqrt{(1-v^2/c^2)}$, and to return to the dimension in the transverse direction the measurements must be divided by this amount.

This factor could not be measured experimentally unless under extreme conditions. In fact, let us see the actual numeric trend of the factor. The following table shows that for the already high speed of 1,000 metres per second (3,600 kilometres per hour), the Lorentz factor remains equal to 1, even with 10 decimal places, and therefore there is (or rather there would be, if the hypothesis corresponded to reality) no observable contraction. For very high speeds Lorentz's factor begins to assume values of some relevance:

Speed in metres / second	Lorentz gamma factor
1,000	1.0000000000
10,000	1.0000000006
100,000	1.0000000556
1,000,000	1.0000055556

For speeds equal to half of that of light, the factor value is about 1.15 and then, for speeds close to the speed of light, it increases rapidly until it becomes infinite for the speed of light:

Speed in metres / second	Lorentz gamma factor
150,000,000	1.154
299,700,000	22.366
299,800,000	27.390
299,900,000	38.733
299,990,000	122.475
299,999,000	387.298

The graph of the Lorentz factor trend is this:

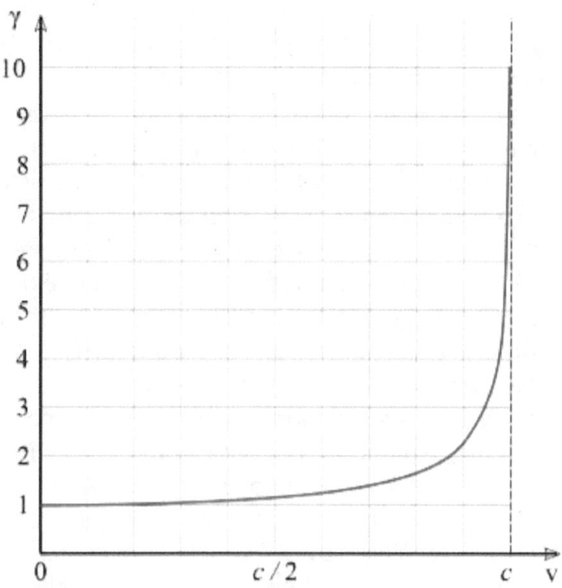

and as we see, the factor remains equal to about 1 for all mechanical speeds, then it grows upward and yields high values for speeds that approximate that of light, and finally it cannot be applied to the speed of light because in this case we would have:

$$\gamma = \frac{1}{\sqrt{1 - \dfrac{c^2}{c^2}}} = \frac{1}{\sqrt{1 - 1}} = \frac{1}{0}$$

and the division of a number by zero does not give any result. Incidentally, this is the reason why later Einstein will say, as is known also by having just heard popular accounts of relativity, that nothing can reach the speed of light. The same consequence applies also to the Lorentz's hypothesis: c is a limit speed that can be approximated but neither reached nor exceeded.

In conclusion, according to the hypothesis of contraction, an object moving for example at the speed of ten thousand metres per second with respect to the ether would be reduced by its longitudinal size of a factor 1/1.0000000006, and its movements and changes would be slowed down of the same factor, obviously not observable and without any practical or technical relevance in all mechanical problems. The hypothesis only serves to make logically consistent the electromagnetic theory, which needs a reference system to be allowed to assume the speed of light as a constant.

In the following, speaking of the contraction hypothesis, for simplicity we shall express ourselves as if the hypothesis were an accepted and consolidated theory, while it is not at all, and it was not even by those who formulated it. This is an *ad hoc* hypothesis that did not satisfy its creators, and was elaborated in the absence of a better explanation for experimental results, and with the expectation that further work around it could have made easier the discovery of empirically verifiable explanations. So, by reading the following, let us remember that we are talking about a hypothetical picture, although for brevity we shall always avoid using the conditional tense in describing the hypothesis. But we shall also remember that we are talking about a work carried out by its authors with the utmost scientific seriousness and with respect for the basic requirements of coherence and logical consistency to which a hypothesis has to conform, however bold and unusual be the behaviour of the physical world described by the theory.

A final clarification and apologies

The most careful readers could have noticed that when deriving Lorentz's factor we allowed ourselves a lack of exactness and a contradiction with what was said earlier: anyway, it was a choice to simplify the argument. In the hypothesis of the ether, as we have seen in paragraph 2.4, it should not be legitimate to represent the path OBO, or OB'O', of the transversal beam with this figure:

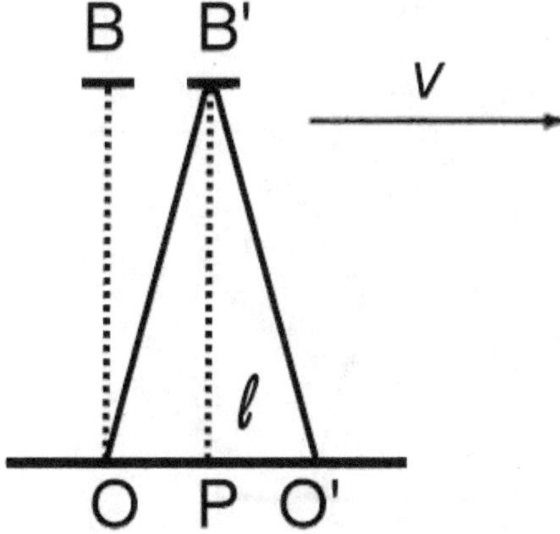

because the beam propagates in the ether and therefore it does not inherit the inertia of the M+M (or similar) device that is moving to the right. So the beam from O should not be reflected by the mirror in B which has moved to B', but it should first go to point B in which there is no mirror anymore, and then go further without ever being reflected. This objection is only valid if we represent the beam as similar to a bullet. But precisely because we are in the hypothesis of the ether, we must not depict the beam like a bullet, but like a wave propagating in concentric circles (more precisely, in concentric spheres) in the fluid that is its condition of existence, as in this figure:

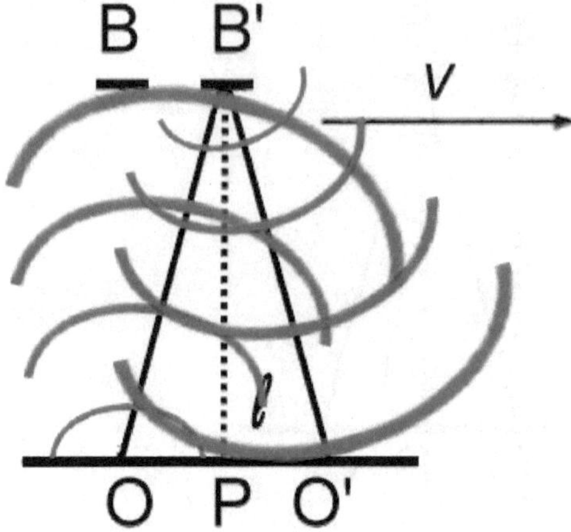

and therefore the beam OB reaches first B' and then O' along twice the length of the hypotenuse of the OB'P triangle, and the argument developed to derive the Lorentz factor is valid and is not spoiled by a logical error, as it could seem at first glance.

2.7   Galilean and Lorentz transformations

Lorentz transformations have consequences on the coordinates of a point in space and time in a Cartesian axis system. The transformation theory is very simple and is described by countless sources, so here we shall give a very concise account of it.

Let us consider a system of Cartesian axes in three dimensions, x, y and z, and consider the coordinate of time, t. The position of a point that moves in this system is defined at every instant by the time value t and by the three spatial coordinates x, y, z.

Now for simplicity, let us only consider the motion along the x-axis, and let us imagine a train moving along it in a straight line, e.g. at 100 metres per second, and a man who is inside the train, at rest or moving. Let's take two systems of Cartesian axes, one having its origin in the station, and the other having origin in the last train car, and call them K and K'. The train moves to the right, along the x-axis, with increasing values of x. So the whole K' system moves right along the x-axis. If at a time t the man sitting has the x, y, z coordinates in the system K (the one having origin in the station), what coordinates will he have in the K' system (the one having origin in the train)?

The answer is very simple, because speed = space/time, and then space = speed × time.

So if the man is in the x position in the system K, to find out where it is in the system K' we have to subtract the space travelled given the displacement of the whole K' system, so we have:

$$x' = x - vt$$

As for the other coordinates, we supposed that the displacement of K' over K takes place along the x-axis, so they do not vary and we have:

$$y' = y$$
$$z' = z$$

And finally, since time is the same for K and K', we have:

$$t' = t$$

This latter consideration is obvious because we are used to considering time to be unique. But now it must be made explicit, for reasons that we shall see very soon.

In the case of two systems K (the station) and K' (the train in motion), where K' moves along the x-axis, we can apply the transformation relationships just seen:

$$x' = x - vt$$
$$y' = y$$
$$z' = z$$
$$t' = t$$

which are called *Galilean transformations*. Galileo indeed described this issue in the famous pages of his *Dialogue Concerning the Two Chief World Systems*, and the rules for transforming the coordinates are consistent with classical mechanics. We also have, of course, the reverse transformation, to apply if we want to get x from x', and here then the speed v changes its sign. The formula can only be:

$$x = x' + vt$$

Obviously if the K' system moved obliquely, we should also transform the y and z coordinates, and the formulas would be somewhat more complex, but for our purposes it is sufficient to consider the simplest case, and consider that a point having coordinates (x, y, z, t) in the K system, will have coordinates $(x'=x-vt, y'=y, z'=z, t'=t)$ in the K' system.

Let us now accept Lorentz's contraction hypothesis, and instead of considering as the origin of the reference systems the train station and the train itself, let us consider as the K system origin any point at rest in the ether, that is, a point of the ether itself or of a celestial body that is at rest with respect to the ether, and as the origin of K' any point in the Earth, being in motion with respect to the ether. The speed at which the two systems move away the one from the other is the unknown one, that of the Earth with respect to the ether. To simplify the problem at best, let us assume again that the x-axis both in K and K' systems is exactly in the direction of the Earth's motion with respect to the ether, so we can we neglect the y and z coordinates.

We know that the lengths of the K' system will be contracted by the Lorentz gamma factor because this explains the failure of the

Michelson and Morley experiment, and we know that the duration of all movements and changes in the K' system will be contracted by the same factor. Therefore, it is necessary to take into account both the two contractions in the transformation formulas between the two systems: not only is it necessary to take into account the contraction of the lengths, but also the contraction of the durations of all phenomena, and thus of the temporal dimension. The lengths, the walls of the houses, the rulers are shorter in K' than in K. Durations, movements, heartbeat, biological clocks in the brains, mechanical clock frequencies are shorter in K' than in K.

So what transformations shall replace Galilean ones? Since we have assumed that motion takes place along the x-axis, y and z coordinates are not affected by the contraction phenomenon, and their transformation is the same as in the Galilean case. Instead, the transformation of the x coordinate must take into account the contraction of the lengths, and that of the t coordinate must take into account the contraction of durations according to the Lorentz gamma factor.

For x, the solution is obvious: just apply the Lorentz factor, so instead of $x'=x-vt$, we shall have $x'=\gamma(x-vt)$, and hence:

$$x' = \frac{x - vt}{\sqrt{1 - \dfrac{v^2}{c^2}}}$$

Regarding time, the transformation does not simply apply the gamma factor, but corresponds to the formula:

$$t' = \frac{t - \dfrac{v}{c^2}x}{\sqrt{1 - \dfrac{v^2}{c^2}}}$$

and therefore a point that has the time coordinate t in the frame represented by the ether, would have the coordinate of time t', determined by applying this formula, in a frame in motion with respect to the ether. The value t' differs from t by a tiny quantity, but not by zero. In order to leave nothing unproven, we shall see below how the transformations can be obtained with all the necessary passages. In particular, the one for the time requires a somewhat laborious procedure.

Now, however, let us only consider that as a whole we have the transformations:

$$x' = \frac{x - vt}{\sqrt{1 - \dfrac{v^2}{c^2}}}$$

$$y' = y$$

$$z' = z$$

$$t' = \frac{t - \dfrac{v}{c^2}x}{\sqrt{1 - \dfrac{v^2}{c^2}}}$$

which are referred to as *Lorentz transformations*.

We know that as far as we are in the world of the Earth or astronomical and mechanical speeds, the Lorentz gamma factor takes values that are not only irrelevant for practical applications, but are not experimentally measurable. The same is true of the numeric values of Lorentz transformations, which as far as we are in the world of earthly or astronomical mechanical speeds, are almost equal to those of the Galilean transformations. So the impact of the Lorentz's hypothesis on the interpretation of the Earth's mechanical world is null. Moreover, it should be borne in mind that even if the speed of light were much lower than it is, and even if the effects of Lorentz's contraction (if it, along with the ether, existed) were noticeable in the mechanical world, the transformations of Lorentz would be useless practically and technically, because to use them in order to correct the coordinates of a system K' that moves over K, one should know which of the two systems moves faster than the other in the ether, and then determine the sign of the displacement speed of the one relative to the other. But we know that no method is available to determine the speed of anything with respect to the ether, and therefore the Lorentz transformations could never be applied correctly: applying them at random we would have 50% chance of doubling a measurement difference, rather than correcting it. The contraction hypothesis has only theoretical use: it serves to explain why the Michelson and Morley experiment and others did not reveal any Earth movement with respect to the ether, and no other purpose. We shall see later that this is not exactly true for high-speed particles phenomena, but for the moment let us be content to understand that for mechanical speeds the situation is the one described here.

Let's take a look at the numerical values of Lorentz's transformations. The following table gives the values with six decimals of x' and t' for the Galilean and Lorentz transformations, on the basis of some values chosen in order to make the numbers easy to interpret, that is:

- let us set in the frame at rest $x = 1,000,000,100$ metres (so at the speed of one metre per second x' becomes exactly $1,000,000,000$ after 100 seconds);

- let us set in the frame at rest $t = 100$ seconds;

and let us calculate the transformed values for speed v between 1 and 1,000,000 metres per second:

Speed m/s	x' Galilean	t' Galilean
	x' Lorentz	**t' Lorentz**
1	1,000,000,000.000000	100
	1,000,000,000.000000	**100**
10	999,999,100.000000	100
	999,999,100.000001	**100**
100	999,990,100.000000	100
	999,990,100,000056	**99.999999**
1,000	999,900,100.000000	100
	999,900,100.005555	**99.999989**
10,000	999,000,100.000000	100
	999,000,100.555000	**99.999889**
100,000	990,000,100.000000	100
	990,000,155.000010	**99.998894**
1,000,000	900,000,100.000000	100
	900,005,100.042223	**99.989444**

The system K' moves along the x-axis at speeds between 1 and 1.000.000 metres per second, and so applying the Galilean transformation the values of x' are decreasing because K' moves away from K, while the value of t always remain the same because the time measurements are equal in K and K'. Instead, by applying Lorentz transformation, Lorentz x' value varies more than Galilean x', while Lorentz t' value is gradually smaller than the Galilean value for both t' and t.

With six decimal places at a speed of 10 metres per second (36 kilometres per hour), a difference of 0.000001 metres is calculated on a total of one billion metres, while it is necessary to reach 100 metres per second to calculate a difference of 0.00001 seconds on 100 seconds. As expected, these values are numerically irrelevant, and the quantities that are calculated through Lorentz transformations are not measurable in the Earth's world and in mechanical phenomena in general.

With this we have the fundamental elements of relativity, in its primitive version, conceived before Einstein to preserve the concept of ether. Einstein's relativity will propose a solution to the M+M experiment problem by explicitly abolishing the concept of ether, but using Lorentz's formulas in the same form, with a different interpretation of their meaning. In both cases, in both versions of relativity, the insignificant numerical values of transformations in the mechanical world allow us to maintain the compatibility of the new theory with classical mechanics: in relation to relativistic mechanics, the classical is not erroneous, but is approximate.

2.8   Formula for speed composition

This detail will be relevant later for relativity in Einstein's version, but we now mention it here to complete the algebraic part of the explanation. In the horizon of relativity, whether it is Lorentz's or Einstein's, the speeds of two moving objects cannot simply be added, but must be computed by taking into account Lorentz's factor. Without considering the problem of contraction, if a train travels at 100 m/sec and a man moves along the train at a speed of 5 m/sec, the speed of the man with respect to the ground is obviously 105 m/sec. Given the speed v of the train and w of the man with respect to the train, the speed of the man relative to the ground will be $u=v+w$. But if the movement took place in the direction of the ether, we could not derive the computed speed u with the mere sum of v and w, but we should take into account the Lorentz factor which determines the contraction of both the train and the man that moves within it, and hence the man's speed w has value w with respect to the train but not to the ground. The formula for computing the composed speed taking into account this factor is:

$$u = \frac{v + w}{1 + \dfrac{vw}{c^2}}.$$

This formula can be obtained with a few passages given the Lorentz transformations. The composite speed is slightly lower than the sum of the speeds, but the difference is not observable, because the divider is just a little over 1 if the speeds v and w are both much smaller than c.

2.9   Length contraction and time dilatation: a recurring misunderstanding

At this point it is good to point out an ambiguity that is generalized in the literature on special relativity, and that derives from imprecise use of the word "time". In all books, "length contraction" and "time dilatation" are used to describe them as characteristics of moving systems (K') with respect to systems at rest (K): and indeed the expression is correct, but only having taken care to avoid terminological ambiguities.

Let us observe the table of the numerical values of the transformations given above in section 2.7: x' values in Lorentz transformations are greater than x' values in Galilean transformations, whereas t' values in Lorentz transformations are lower than the Galilean value, which always remains the same. This suggests the idea that what is gained in time is lost in space, or vice versa. But this is not exactly the meaning of Lorentz hypothesis. In all moving systems with respect to the ether, the measurement of the speed of light must return the same value, c, otherwise the measurement differences would allow the Earth's speed with respect to the ether to be known. Therefore, since light speed is measured with devices that are more complex than rulers and chronometers, but conceptually are equivalent to them, in all moving systems with respect to the ether the rulers have to contract by the gamma factor, and hence also the duration of phenomena and the movement of chronometers if it does not remain constant, must contract, that is, slow down, by the same gamma factor. If this were not the case, the speed of light would have different values in moving systems compared to the ether at different speeds, because the relationship between spaces and times would be different. Instead, given the failure of any experiment to reveal the Earth's speed with respect to the ether, in all systems where we measure the speed of light as a ratio between a space segment Δs and a duration Δt we must always have:

$$c = \frac{\Delta s}{\Delta t} = \frac{\Delta s'}{\Delta t'} = \frac{\gamma(\Delta s)}{\gamma(\Delta t)}.$$

2. Ether, electromagnetic theory and Lorentz's hypothesis

To avoid the ambiguities associated with this issue, we shall use precise terminology and say that we have *points*, and we have *space segments* or *lengths*, and that these are contained in *space*. We shall say that we have *instants* and we have *time segments*, or *durations*, and that these are contained in *time*. So in a system that moves with respect to the ether we have a contraction of lengths and a contraction of durations, or, if we want to express it in the mutual way, we have a dilatation of space and a dilatation of time. Given any K' system in motion with respect to K, which is assumed at rest with respect to the ether, a one metre long ruler in K indicates a length of less than one metre in K', and a clock that marks an hour in K indicates a duration of less than one hour in K'. Hence lengths and durations are both contracted in K', or, that is to say the same thing, both space and time are dilated in K': in the following we shall use this unambiguous way of expression without introducing improper suggestions. The treatises about relativity (I do not know if there are any exceptions, but I have never encountered one that talks about this issue differently) usually express themselves by putting together the terms *lengths contraction* and *time dilatation* in the systems in motion. These expressions are correct literally, as we have just observed, but it can puzzle readers by suggesting an erroneous interpretation, as if these two terms were privileged with respect to the other two in play, and does not help at all to understand the theory in the interpretation that Einstein worked out. The confusion between contraction and dilatation, invoked without specifying whether it is about lengths or space, durations or time, suggests that there is a sort of balance between the time gained and the lost space, or vice versa. This does not make sense considering Lorentz's contraction, but has a particular sense that emerges considering the form of the Lorentz transformations: this will be discussed later, when it will be time to talk about the concept of space-time interval in the theory of special relativity.

2.10   Relativity of simultaneity and time measurements

Where the Lorentz factor would apply, simultaneous events would also have different time coordinates. In general, a phase shift of the time coordinates corresponding to the transformation for the t coordinate would occur.

Let us suppose now that there are inhabitants of Mars who measure time in a way similar to that of Earth's inhabitants, and that Martians and Earth dwellers have identical chronometers. The chronometers of the dwellers of both planets mark the time exactly the same way as they are at rest in the ether, and of course they are subject to the same contraction of time durations when they are in motion, because Lorentz's hypothesis is a universal hypothesis. Let us suppose that at some instant Mars is at rest with respect to the ether (during the year of Mars this could happen), while the Earth moves with respect to the ether and moves away from Mars. Suppose that the inhabitants of Mars and Earth observe two phenomena that happen on a third celestial body, for example, two sunspots that appear one after the other. Given the phenomenon of general contraction of lengths and durations, at this time of the year the Earth's running clocks are slower than the identical chronometers at rest on Mars. Therefore, if a Mars and an Earth inhabitant started their stopwatch at the time of the first sunspot, assigning time=zero to this event, and then read the stopwatch when the second spot appears, they would read two different values as the time of the second event.

For the sake of completeness, let us imagine that Martians and Earth dwellers both take into account the fact that the light has a finite speed, and that in order to attribute a time value to solar phenomena, the delay with which their image arrives on both Mars and Earth is taken correctly into account. On the Earth the average value of this time, as everyone knows, is about 500 seconds, while on Mars it is about 800 seconds. In our example of sunspots, both the inhabitant of Mars and that of Earth take into account the delay with which they see the spots, and both assign to the appearance of the first spot the time zero, and to the second the time t measured with their stopwatch. Due to Lorentz's contraction, the time t marked by the Earth's stopwatch is less than the time t read on Mars. Given the hypothesis of contraction of durations, the fact that the two observers measure different times with their chronometers is a perfectly logical consequence, however, from the point of view of a third observer who could know all the factors involved (sunspots, time measurements made on Mars and time measurements made on Earth) the result would be unacceptable, because this third observer would say: two sunspots were born on the Sun, and therefore the time gap between them is one and must be objective and unambiguous. The third observer in this way judges that the same event, the formation of the second solar spot, receives different time coordinates on the two planets, and so realizes that events he considers simultaneous could be considered not as such on Earth and Mars.

We know already, from the considerations made in the first chapter, that every observer who constructs an interpretation of the phenomena which he has experienced, puts them inside a unique time, because he cannot do otherwise: each one organizes the phenomena in a single line of time. Every living man perceives phenomena by accumulating them in memory, but when he feels the need to organize them, he reports them on one unique line representing time; every social system of men in mutual communication defines its own calendar, with which socially known and shared phenomena are referred to on a single socially shared line of time. When two social systems which were isolated in the past come into communication, they soon make their calendars compatible and mutually transformable. But systems (and here we talk about social systems, made of humans or any kind of thinkers, even Martians, provided they are conscious similar to humans) which do not communicate with each other, refer the phenomena each to their own line of time, and there is never reason to think that time measurements are consistent between isolated and non-communicating systems.

In the hypothesis of Lorentz's contraction, to this general relativity of time — the one that causes the temporal references of non-communicating systems to be indeterminate one with respect to the other — is added a second relativity factor, determined by the contraction, which causes identical clocks to mark different times in relation to their current motion state with respect to the ether.

The inhabitants of Mars and those of the Earth in our example would give a different time t for a phenomenon, the appearance of the second solar spot, while from the point of view of a third observer who considers the whole of the problem, this phenomenon is the same, and so it should have the same value t of time in both systems, after the time of the first solar spot has been assumed as zero time. But until the inhabitants of Mars and Earth proceed without knowing anything about each other, they both make their measure of space and time without knowing anything about Lorentz's contraction, which is not perceived. Knowing only their measurements of space and time, the inhabitants of Mars and Earth would measure a time t for the instant of the second solar spot that would be different in the two cases, that would be relative to the Mars system or the Earth system. There would be no reason to consider one of the two in error: each would perform the same measuring procedures correctly.

But if the inhabitants of Mars and of the Earth came into communication with each other, or if their operations were taken into account by a third observer, then the need to relate the instant of the second event to a given unique time system would arise. It could be that of Mars, that of the Earth, or a third system: in any case the choice would be conventional. If it were possible to locate who is stationary and who moves with respect to the ether, a thing never happened, it would be an obvious practical choice to assume the timing of clocks at rest in the ether as the fundamental time values: but this would always be a conventional choice, logically equivalent to the others that could be done. In any case, Lorentz's transformations would be the key to calculate the common value of time t for each phenomenon, whatever be the choice made for the common time frame.

2.11   Derivation of Lorentz transformations

Above we have outlined the Lorentz transformations without showing why they have exactly the form they have. Since it is enough to know what they mean, and the reader does not need to know how they can be derived, those who want can jump over this section without prejudice to the understanding of all that follows. On the one hand, I strongly advise readers to invest in the labour that is needed to follow the arguments of the next pages: this will be rewarded by the acquisition of a strong sense of familiarity with Lorentz's theory, and also the aspects of Einstein's relativity already known in a generic way will begin to take a definite form in the mind of those who assimilate this argument. But if readers have understood all the above, and they feel that the tightened attention that is needed to read this section is distracting, then they can read further now and come back to this argument later.

To derive Lorentz's transformations, I reworked the technique used for the same purpose by Henri Bergson in his treatise on Special Relativity in 1922. We shall discuss later this important treatise, which is distinguished from relativistic literature for the particular interpretation it gives the work of Einstein, but before dealing with Einstein's relativity it presents Lorentz's hypothesis with exemplary clarity, and therefore it is still a useful tool. Lorentz's hypothesis allows us to understand the concepts of relativity by remaining in the perfectly intuitive mental framework of classical physics, and therefore it is useful to assimilate it fully before reading Einstein's.

As an introduction, let us begin to realize only the transformation for the x coordinate, which is the easiest to derive. We as always have a fixed system of origin O stationary with respect to the ether, and a moving system with origin O' that moves at speed v in the direction of x which is longitudinal with respect to the motion of the Earth in the ether. At time t_0=zero, let us say that the moving system coincides with the fixed system and let us set zero time in both systems: $t_0=t_0'=0$.

At a given instant t, while the moving system is in motion, x is the coordinate of a point located in position M. If t is the time in the fixed system and x is the position on the x-axis in the fixed system, we must find the corresponding t' and x' values in the moving system subject to contraction of distances and durations.

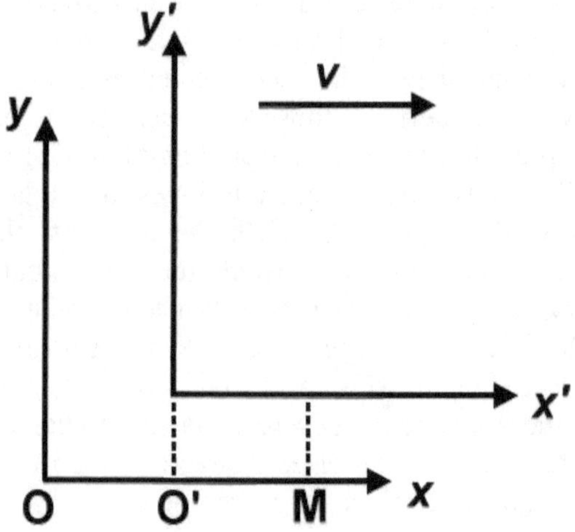

Finding x' is easy. We have in the fixed system OM=OO'+O'M where OM=x, OO'=vt, that is the space travelled by the moving system in time t, and O'M=x'/γ because it is assumed that O'M is contracted. Let us remember that:

$$\gamma = \frac{1}{\sqrt{1 - \dfrac{v^2}{c^2}}}$$

then γ is a greater quantity than zero, and therefore O'M, which is contracted, must be divided by γ.

We therefore have in the fixed system:

$$x = vt + x'/\gamma$$

hence follows soon:

$$x' = \gamma(x - vt)\dots(1)$$

which is the first transformation.

Instead, to find t' we need a more extensive consideration and the development of an argument with which we also introduce a kind of reasoning that later will be used extensively in dealing with Einstein's relativity.

2. Ether, electromagnetic theory and Lorentz's hypothesis

Let us remember that in Lorentz's hypothesis anyone, at rest or in motion with respect to the ether, always measures the same value for the speed of light, and that this fact defines the problem of ether: if the speed of light had different values, its measure would allow us to find the Earth's speed with respect to the ether. But that does not happen. So, as we know, Lorentz's hypothesis explains the constancy of the speed of light through the contraction of lengths and durations in all things, and therefore also in measuring equipment. Having said this, we imagine that there is a system formed by a rod OA mounted on the Earth and in motion at speed v together with the Earth, i.e. v is the speed with which the Earth is currently moving in the ether, and we imagine that there is an external observer that we call Z who is at rest with respect to the ether (that is, on some planet that is at rest) and knows the state of motion of the OA system. The OA system could be, say, a 1000 km long straight highway at whose extremities there are people who can communicate through light or radio signals. Z sees the Earth from his or her observation point at rest and sees the OA system on it that gets farther to the right moving in the ether:

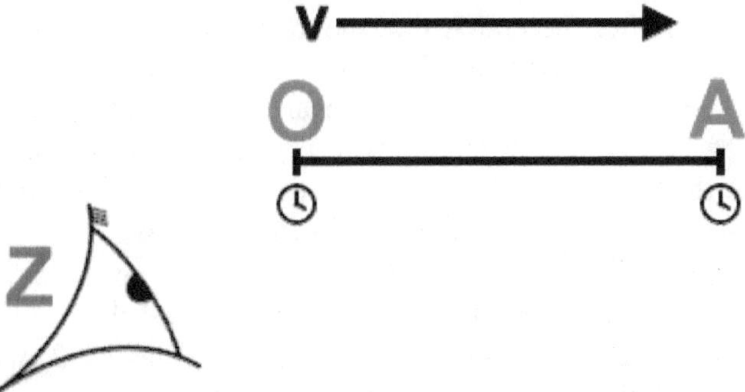

Let's suppose now that there are two clocks at the ends of the system in motion that need to be synchronized, in order to mark the same time. To get synchronization, people in O and A agree to perform this procedure:

- O resets his clock and sends a light signal to A. The zero of O is called "temporary zero".

- A receives the signal and reflects it towards O and while doing this, exactly at that instant, resets his clock.

- O receives the return signal, reads the time t marked by his clock, and adjusts his clock to $t/2$. In fact, O assumes that the time required for the journey forth and the journey back of the signal is the same, and therefore assumes that, if for example the total return time of the signal is 10 seconds, the events would be:

Event	The clock in O marks	The clock in A marks
The signal starts from O	0	nothing
The signal arrives at A	5	0
The signal comes back to O	5+5=10	5
O corrects his clock	10/2=5	5

If O and A had performed this procedure being at rest in the ether, they would have good reason to believe the procedure was correct. But if the moving system were seen by the observer Z who knows to be at rest with regard to the ether, Z would say that the OA and AO journeys are different, because in the journey forth the beam started in O has to run after the end A that moves to the right and escapes, while the beam reflected from A returning takes less time to reach the end O because the end O moves towards it. According to Z, the OA distance is travelled at the speed $c-v$, while the distance AO is travelled at the speed $c+v$. So according to Z it is not right to adjust the clock in O on $t/2$ to mark the same time as clock A.

If the distance OA is equal to l, from the point of view of the observers in motion in the ether, the beam will travel the distance $2l$ on the roundtrip and the total time, which O divides by 2, will be $2l/c$. Because of the length and duration contractions, the speed of light is always c, and always on account of the contraction of durations the O clock at the time of receiving the reflected beam will mark the value $2l/c$. But the observer standing at rest in the ether, seeing the motion of the beam from O to A and from A to O, judges that the distance travelled by the beam is not $2l$, but $2l$ contracted according to the gamma factor, and hence:

$$\frac{2l}{\sqrt{1-\dfrac{v^2}{c^2}}}.$$

Always according to the observer Z at rest, if the clock of O wants to mark the time in the same way as the watch that Z wears on his or her wrist, it should not mark *2l/c* but:

$$\frac{2l}{c} \times \frac{1}{\sqrt{1-\dfrac{v^2}{c^2}}}.$$

But the clock in motion is late and only indicates *2l/c*, because the phenomena of the moving system are contracted and there time is dilated: the evolution of phenomena in time flows more slowly. Given the same interval between two events, the moving system clock counts less seconds, and therefore a second lasts more. Precisely, the second marked by a clock of the moving system with respect to the ether has the duration of:

$$\frac{1}{\sqrt{1-\dfrac{v^2}{c^2}}}$$

seconds marked by a clock at rest.

But the inhabitants of the Earth and everybody else who is in motion with respect to the ether do not realize it. Let us calculate exactly all the space and time measurement differences. We know that the time needed for the beam roundtrip used to synchronize the clocks in the moving system is *2l/c* and that this is the value indicated by the O clock when the return beam reaches it. Then, if at the time of the signal emission a temporary zero was marked in the position indicated at that time by the clock hand, at the end of the procedure, the definite zero time is assigned by O to the position *l/c* (the half of the time detected) on the dial of his clock; we call M this position, and O believes that it corresponds to the zero marked by the clock in A. But the observer at rest knows that the signal travelled at speed *c-v* when going forth and at *c+v* when returning, and therefore knows that if the zero of the clock in O must correspond correctly to the zero of the clock in A, the range *2l/c* should not be divided into two equal parts but in two parts proportional to *c+v* and *c-v*. If we call *x* the first of these two parts, we shall have:

$$\frac{x}{\dfrac{2l}{c}-x} = \frac{c+v}{c-v}$$

hence:

$$x = \frac{l}{c} + \frac{lv}{c^2}$$

which means that according to the observer at rest the point M on the clock dial where A has set the definite zero is too close to the provisional zero of lv/c^2 seconds, and then to place in the same time coordinate the zero of the two clocks O and A, the zero of clock A should be brought back by lv/c^2 seconds. So the clock in A is always late for lv/c^2 seconds with respect to the time it should mark. Let's call t the time of the motionless Z clock in the ether and t' the time of the moving clock. When the clock hand in A marks t', according to the resting observer Z it should mark

$$t + \frac{lv}{c^2}$$

in order to mark the same time as the clock in O.

That said, what would happen if two physicists who are in O and A respectively wanted to measure the speed of light using clocks synchronized with their criterion, and used the first clock to detect the starting time of a light beam and the second to detect the arrival time?

Clocks give such indications so anyone who considers them synchronized believes that the time required for the OA path going forth of a beam is equal to the time required for the AO return path. Therefore, the two physicists will think that the time required for the journey from O to A detected by the two clocks in O and A is equal to half the time of the complete roundtrip in the two senses measured only by the clock in O. However, we know that the duration of the journey forth and back if determined by the clock in O is always the same, regardless of the speed of the moving system. The same is true for the duration of the journey in one way detected by the new procedure that uses two clocks: and for this reason, the constancy of the speed of light will continue to be observed. However, the observer at rest in the ether will follow the occurrence of the events one after the other and to him it will be clear that the distance travelled by the beam from O to A is proportional to the distance travelled from A to O according to the $c+v$ to $c-v$ ratio and hence that the two paths are not the same. He will find that since the zero of the second clock does not agree with the zero of the first, it happens that to the moving observers times appear the same following the indications of their clocks, while in reality the two times are in the ratio of $c+v$ to $c-v$. The resting observer Z concludes that there is an error in distance and error in time measurement in the moving system, but will also conclude that the two errors are compensated because they are the consequence of the error introduced earlier, when moving clocks were synchronized. For this reason, in the moving system (which from now on we shall call S') it is possible to obtain an equal value for the speed of light, either by measuring time by one clock or by using two clocks distanced from each other. Observers residing in the moving system will find that the second experiment confirms the first, while the Z observer at rest will realize that to correct all the time measurements made in the moving system S' one should work distinct corrections for all existing synchronous clocks in S', which in our case until now are two. He already knew that running clocks are slower than those at rest. Now he knows more, i.e. he knows than the moving clocks arranged along the direction of motion delay one with respect to the other, if they are synchronized with light beams. Suppose now that the moving system S' is just a kind of a clone of S system at rest, as if the Earth gave birth to its own double, like a huge amoeba. Suppose that the second system S' has been generated and separated from S when a clock C_0 in system S marked zero. In the new system there are clocks that clone all the clocks in the first, and at the time of the birth of S' the clocks C'_0 and C_0, of which C'_0 is the clone, and so on for all the clocks C'_n and C_n, mark the same time. Now in the system S' let us consider a clock C'_1 that is in such position so that the straight line between C'_0 and C'_1 is parallel to the direction of motion, and let us call l the distance between C'_0 and C'_1. When clock C'_1 indicates time t', the observer at rest observes with good reason that since the clock C'_1 is delayed with respect to the C'_0 clock of the moving system of a range of lv/c^2 seconds, in reality in the S' system has elapsed a time equal to $t'+lv/c^2$ seconds. He had already observed the slowdown in clocks and in all the phenomena due to motion with respect to ether, and he knew already that every second in the system S' equals:

$$\frac{1}{\sqrt{1-\dfrac{v^2}{c^2}}}$$

"true" seconds,

or better seconds measured at rest with respect to the ether. Thus the observer at rest is able to find that if the clock C'_1 gives the indication t', the time really elapsed is:

$$\frac{1}{\sqrt{1-\dfrac{v^2}{c^2}}}\left(t' + \frac{lv}{c^2}\right) \quad (2)$$

and if he consulted for confirmation one of his clocks at rest, he would always read a time t that indicates this value. Let us remember this formula (2) because we shall have to consider it again at the end of the argument.

But even before determining the correction necessary to transform time t' in its time t, the observer at rest would realize a mistake made in the moving system in judgments about the simultaneity of events. This is the result of the synchronization procedure of moving clocks. Let us suppose that in the moving system the line in the direction from C'_0 to C'_1 is prolonged for many kilometres, and along it there is a large number of moving clocks C'_0, C'_1, C'_2 and so on, all separated by lengths equals to l. When the system S' coincided with S and was thus motionless in the ether, the optical signals went back and forth between the pairs of subsequent clocks, running equal spaces in both senses. So after the synchronization operation when two or more clocks synchronized with light signals read the same time t, the t indication was given by the clocks actually at the same instant. At a certain moment the S' system has detached itself from S as a result of cloning, and the people inside S', who do not know they are moving, allowed the clocks C'_0, C'_1, C'_2, etc., to continue running in the state they were in; people in the moving system so far reasonably think that when the clocks indicate the same time value, the events occurring near the clocks are truly simultaneous. In fact, due to motion with respect to the ether, all the clocks begin to delay, but moving at the same speed all delay by the same factor. But suppose that later people in the moving system repeat the synchronization procedure of all their clocks: during the new synchronization the observer Z at rest sees the signals going forth from C'_0 to C'_1, from C'_1 to C'_2 and so on and knows that they travel longer spaces than those returning from C'_1 to C'_0, C'_2 to C'_1 and so on, and then realizes that to have correct simultaneous indications, when the clocks indicate the same time values it would be necessary to bring back the zero of clock C'_1 of *lv/c²*, to bring back the zero of clock C'_2 of *2lv/c²*, the zero of clock C'_3 of *3lv/c²*, and so on. The simultaneity in the motion system has become nominal due to the synchronization procedure: or at least, so thinks the observer Z.

So even if the system S' is an identical clone to the system S, because now it moves at speed *v*, it is undergoing these changes with respect to S:

1)   All lengths in S' have contracted in the direction of motion. The new lengths are proportional to the old ones in the relationship:

$$\sqrt{1 - \frac{v^2}{c^2}} \, .$$

Reciprocally, space in the system S' has dilated with the inverse factor in the direction of motion.

2)   The durations of all phenomena in S' have been shortened. The new durations are proportional to the old ones in the relationship:

$$\sqrt{1 - \frac{v^2}{c^2}}$$

Reciprocally, time in the S' system has dilated with the inverse factor. Every second of time in the S' system has the value:

$$\frac{1}{\sqrt{1 - \frac{v^2}{c^2}}}$$

"true" seconds of the system at rest.

3)   If the clocks have been synchronized with optical signals along the motion line, the simultaneities of the system S have been generally converted into succession in the system S'. If synchronization occurs along any line perpendicular to that of

motion, events that are simultaneous in S remain simultaneous in S'. The events in the direction of motion that would be simultaneous in S, in S' are separated by lv/c^2 seconds of the system S'.

So even if S' were a clone of S perfectly identical to S for any physical property, S' would have spatial and temporal properties differing from S. But the inhabitants of S' would have no notion of it, and only an observer who is immobile in ether and studies the situation of both systems is able to realize it.

At this point let us imagine that there are two scientists, both observers in the two systems, which we call Dr. Static and Dr. Dynamic, who are able to communicate with each other via radio. Dr. Static has observed the events and could therefore tell Dr. Dynamic: "After you've started, your spaces have deformed, your times have inflated and your clocks have given up marking at the same time. The measurements are incorrect according to exact computable amounts, so it is now your duty to apply the formulas we know and correct your measures to re-establish the truth." But Dr. Dynamic would answer: "I shall not apply any corrections, because if I did apply your formulas to my world, it would become incoherent from a practical and theoretical point of view. You say that my lengths have contracted: but you know that also the one-metre ruler I use to measure them has contracted, so I do not have to do anything else than go on measuring things as I did before. And the same is true for time measurements. You say that when the hand of my clock has advanced a second, the hand of yours has advanced more than a second. But in my world as in yours the rotation of the Earth occurs in a day made of 86,400 seconds, and life takes place as before. And finally, do you say that the simultaneity of certain events has been converted into succession? That that three lined clocks C'_1, C'_2, C'_3 indicate the same time at instants which you consider different? But in those moments which my clocks indicate as simultaneous, events occur that in my system S' are legitimately considered simultaneous, and so for me the simplest choice is to keep on considering them simultaneous: because doing so I do not have to do anything, while if I wanted to persist in making my time measurements equal to those of your system at rest, I should correct all the measurements of time I would later make in my moving system: my clocks would no more mark the time useful for us dwellers of S' in motion, and I should continually correct them. By transforming into succession what seems to me simultaneity, I would build an incoherent world, and it would not even be anymore a clone perfectly identical to your world. If I just do nothing, all the things in my world and all the relationships between things will continue to run according to the same physical laws of your world. My world has undergone major changes in spatial and temporal measurements and relationships, but I do not need to take it into account."

There is, however, a case in which Dr. Static and Dr. Dynamic must take into account the transformation that the system in motion undergoes. Dr. Dynamic explains the case in this way: "The case is if we wanted to build a unified mathematical representation of our systems, so we would also build a unified mathematical representation of all systems in motion or at rest with respect to the ether that may exist in the Universe. In order to create the representation that allows us to know the relationship of each thing with each other, we should define every point of the Universe by the distances x, y, z from three orthogonal planes that we should lay as a basis at rest for each measure, and which intersect defining three OX, OY, OZ axes. These axes OX, OY, OZ, which will be the preferred frame on which to base each measure, will be the only really and non-conventionally axes at rest, and therefore have to be defined with respect to the system S that is at rest in the ether. In the motion system in which I live all my measurements will be made in reference to three other axes O'X', O'Y', O'Z', which will be in motion along with my system. In my system all points will be defined by the distances x', y', z' measured from the three planes orthogonal to O'X', O'Y' and O'Z'. But since the global representation of our two systems and of every other system in the Universe must be based on the three planes OX, OY, OZ at rest in the ether, it is necessary to find the form of the equations with which, given the coordinates of a point x', y', z' in any moving system, we can compute the corresponding coordinates x, y, z in the one at rest. The information we have is sufficient to derive these equations.

To simplify the problem, we assume that the movable axes O'X', O'Y', O'Z' are coincident with the corresponding fixed axes before the dissociation of the two worlds S and S', and we shall also assume that OX and O'X' are oriented in the direction of the movement of S'.

2. Ether, electromagnetic theory and Lorentz's hypothesis

Taking these assumptions, it is clear that the Z'O'X' and X'O'Y' planes simply slide on the ZOX and XOY planes respectively, and that these planes therefore always coincide and therefore y and y' are always equal, as well as z and z'. It only remains to calculate x. If you measure a time t' on a clock at x', y', z', after the instant in which O' separates from O, the distance of this point from the ZOY plane will be equal to $x'+ vt'$. But as a result of the contraction of the moving systems, the length $x'+vt'$ will not coincide with the x in the system at rest, but with:

$$x \sqrt{1 - \frac{v^2}{c^2}}$$

and hence the value of x in the fixed system is:

$$x = \frac{1}{\sqrt{1 - \frac{v^2}{c^2}}} (x' + vt').$$

This solves the problem with space measurements. But we must not forget that even the time t' elapsed in the moving system and indicated by the moving clock at point x', y', z' is different from the time t of the system at rest. When the moving clock has indicated t', the t time shown in the fixed system is:

$$\frac{1}{\sqrt{1 - \frac{v^2}{c^2}}} (t' + \frac{vx'}{c^2})$$

because we have seen above, in formula (2), that this is the correction of the time measurement performed by a clock that has shifted for the length l at speed v. Since l is a length measured in the moving system, we can replace it with x'."

Here ends the explanation by Dr. Dynamic, who in this way would find the form of the Lorentz transformations for x, y, z and t:

$$x = \frac{x' + vt'}{\sqrt{1 - \frac{v^2}{c^2}}}$$

$$y = y'$$

$$z = z'$$

$$t = \frac{t' + \dfrac{v}{c^2}x'}{\sqrt{1 - \dfrac{v^2}{c^2}}}$$

And once found these relationships, just solve them by x' and t' to get the form in which the transformations are usually presented, which we already knew:

$$x' = \frac{x - vt}{\sqrt{1 - \dfrac{v^2}{c^2}}}$$

$$y' = y$$

$$z' = z$$

$$t' = \frac{t - \dfrac{v}{c^2}x}{\sqrt{1 - \dfrac{v^2}{c^2}}}$$

And with that, we said all that was necessary to know about Lorentz's hypothesis. We can move to relativity in Einstein's version.

3.   *Einstein's solution*

3.1   *Two Einstein's postulates*

At this point, we know the premises necessary to understand the solution proposed by Einstein to the problem, in which we shall recognize concepts that have become familiar to us, but which will not be easy. The sections from here up to 3.9 will be the most difficult, and eventually section 3.10 will provide a summary that can help readers to arrange their ideas. Einstein's theory is to be assimilated with mental elasticity, trying to get a clear idea of what Einstein tries to express and why he expresses it in a certain way, but without pretending to find soon a satisfactory and unambiguous interpretation of what he says. What we are now explaining cannot be described with the linearity with which Lorentz's theory can be outlined.

We can immediately tell in a few words the central idea of special relativity, and thanks to what we have seen so far the generic meaning of these words will be clear: the central idea is to apply Lorentz's formulas to any rectilinear uniform motion without referring to the ether, removing the concept of the ether from our conceptual horizon, and assuming that electromagnetic waves propagate in the vacuum from the bodies that emit them. A simple idea, but not easy to justify and make plausible: what can authorize us to apply Lorentz's math, if we no longer assume the ether, and therefore not even the contraction of durations and lengths in motion with respect to the ether?

To understand the solution proposed by Einstein we shall follow the text of the 1905 paper *On the electrodynamics of moving bodies*, quoting some parts also in German for those who want to evaluate nuances of meaning, and we shall integrate it with some reference to the popular disclosure of relativity published by Einstein in 1916 and to other texts. From now on the paper of 1905 will always be indicated as ED1905. In the 1905 paper Einstein uses for the speed of light the capital letter V: to facilitate readability, this expression is now replaced by the usual small c, entered into use a few years later. Einstein's text also uses the expressions K' to indicate the moving system with respect to K, and x', y', z', t' for the coordinates of the moving system with the meaning we know, but with the notation: k

(small) instead of K' and the Greek letters ξ, η, ζ, τ instead of x', y', z', t'. By quoting the text for practicality I replaced this notation too with the one that became usual. The paper contains a brief premise and then ten chapters, for a total of 31 pages in the original edition.

What we have seen so far allows us to understand why Einstein's theory of special relativity has a title that refers to "electrodynamics" and not to problems of geometry of motion or to space and time. Electromagnetic theory, led to a complete and unified form by Maxwell, proposes physical laws in which light speed c is a constant. But each speed presupposes the spatial frame in which it is defined; therefore, until the ether hypothesis was adopted, the speed of light and of all electromagnetic waves was defined as propagation speed in the ether. But lacking the notion of ether, electromagnetic theory goes into a paradoxical condition: it records new experimental and technological successes every day — just mention the name of Marconi to be sure of it — and at the same time it does not know how the speed of light may be a constant. If electromagnetic waves are emitted by any kind of body, lamp, or radio transmitter, their speed will be c relative to the emitting body, and will be $c+v$ or $c-v$ if the emitting body moves at speed v with respect to anything else. Thus the expression c cannot enter into the formulation of universal physical laws: light will always have a certain relative speed with respect to the frame that serves to measure it, and therefore the form of the most useful laws of Maxwell appears to contain an illogical and unsustainable assertion. That is why the central problem is "the electrodynamics of moving bodies". As a result, Einstein's paper begins with these words[13]:

> *Daß die Elektrodynamik Maxwells — wie dieselbe gegenwärtig aufgefaßt zu werden pflegt — in ihrer Anwendung auf bewegte Körper zu Asymmetrien führt, welche den Phänomenen nicht anzuhaften scheinen, ist bekannt.*

> It is well known that Maxwell's electrodynamics — as usually understood at present — when applied to moving bodies, leads to asymmetries that do not seem to match with the phenomena.

---

[13] ED1905, Foreword.

3. Einstein's solution

Shortly after mentioning a more detailed circumstance we shall deal with in the following, Einstein mentions experiments that have suggested abandoning the notion of ether[14]:

> *... die mißlungenen Versuche, eine Bewegung der Erde relativ zum "Lichtmedium" zu konstatieren, führen zu der Vermutung, daß dem Begriffe der absoluten Ruhe nicht nur in der Mechanik, sondern auch in der Elektrodynamik keine Eigenschaften der Erscheinungen entsprechen, sondern daß vielmehr für alle Koordinatensysteme, für welche die mechanischen Gleichungen gelten, auch die gleichen elektrodynamischen und optischen Gesetze gelten ...*

> ... the failure of attempts to detect a motion of the Earth relative to the "light medium", lead to the conjecture that not only in mechanics, but in electrodynamics as well, the phenomena do not have any properties corresponding to the concept of absolute rest, but that in all coordinate systems in which the mechanical equations are valid, also the same electrodynamic and optical laws are valid ...

Immediately after, Einstein announces the two hypotheses, or suppositions, or postulates, which he intends to develop. The first hypothesis is what has just been said, that is, "in all coordinate systems in which the mechanical equations are valid, also the same electrodynamic and optical laws are valid." Einstein assumes it as a postulate[15]:

> *Wir wollen diese Vermutung (deren Inhalt im folgenden "Prinzip der Relativität" genannt werden wird) zur Voraussetzung erheben ...*

> We shall raise this conjecture (whose content will be called "the principle of relativity" hereafter) to the status of a postulate ...

and then he adds the second hypothesis, which is this[16]:

> *... und außerdem die mit ihm nur scheinbar unverträgliche Voraussetzung einführen, daß sich das Licht im leeren Raume stets mit einer bestimmten, vom Bewegungszustande des emittierenden Körpers unabhängigen Geschwindigkeit c fortpflanze.*

> ... and we shall introduce, in addition, the postulate, only seemingly incompatible with the former one, that in empty space light is always propagated with a definite speed c, which is independent of the state of

---

[14] ED1905, Foreword.

[15] ED1905, Foreword.

[16] ED1905, Foreword.

3. Einstein's solution

motion of the emitting body.

The first of the two assumptions expresses nothing more than a principle of methodological coherence: since modern physics is built on the law of inertia, according to which the state of rest and uniform rectilinear motion cannot be distinguished qualitatively, but are always relative, it is necessary to find the way to reconcile Maxwell's electrodynamic laws with this principle. This extension of the law of inertia of Galileo and Newton to electromagnetic theory and to every physical law generally takes the name of "principle of relativity", and its formulation is not to be ascribed to these words of Einstein, but to Poincaré, which had given similar statements between 1900 and 1905, dealing with Lorentz's hypothesis, to which he attributed importance and plausibility, and about which he wanted to find the way to overcome the exception with respect to the principle of relativity. Although Poincaré was not quoted, the specialists of that time immediately recognized that Einstein's attitude was inspired by Poincaré, who died in 1912, and had time to acknowledge Einstein's theory, which he did not accept, with an explicit statement[17].

The second of the two suppositions is that light always propagates at speed c, whatever the chosen frame, and hence without being added to the speed of the emitting body. Einstein knows that the second assumption, or postulate, is difficult to accept, and recognizes that at first glance it appears incompatible with the former. Having studied Lorentz's hypothesis, however, we are prepared for the idea that with some further explanation we could accept this postulate: in fact, given the contraction of distances and durations, or the dilatation of space and time in the case of motion compared to the ether, in Lorentz's theory light speed measurements would always return the same value, because the effect of the observer's motion with respect to the ether is cancelled. Einstein's postulate corresponds in a certain sense to data from experience, for those who know the issue of the ether and also the hypothesis of Lorentz.

---

[17] Poincaré's position is briefly but completely accounted in Canales 2015, Chapter 6. Canales also provides extensive bibliographic information.

3. Einstein's solution

However, the sense of the constancy of the speed of light in the conceptual framework of Lorentz's hypothesis requires both the ether and the contraction of lengths and durations. The speed of light is relative to the ether, is constant, and the measure always finds the value c even if the observer is moving in the ether, due to the space and time contraction of measuring instruments (to be rigorous, we know that only lengths contraction is necessary, however if we want we can add the hypothesis of durations contraction too). If we remove any of the two factors, either ether or contraction, it becomes very difficult to imagine that the speed of light can be an absolute constant and not a magnitude dependent on the speed of the emitting body. Einstein is aware of that, and will try to make plausible the assertion with a different theory than Lorentz's.

Let us forget Lorentz's hypothesis, and let us schematize the problem in an elementary situation. Let us imagine a three-lane highway, and a truck that travels at 90 km/h with respect to the ground, a car travelling at 120 km/h, and a motorbike travelling at 150 km/h. We shall say that the car travels at 30 km/h with respect to the truck, and the motorbike at 30 km/h with respect to the car and at 60 km/h with respect to the truck. Keeping the time and space references stable, we cannot give any sense to the statement, that somebody could pronounce, that the speed of the motorbike is 150 km/h compared to all three reference systems: ground, truck and car. Now suppose the motorbike switches on its front lights, and suppose there is no ether and that the headlamp light is propagated at a certain speed with respect to the lamps or the LEDs that emit it: we shall say that the light propagates from them at speed c with respect to the motorbike and (after transforming c into hourly kilometres) we shall say that it moves at speed $c+30$ with respect to the truck, at $c+60$ compared to the car and at $c+150$ compared to the ground.

But to accept this way of representing the problem, it would be necessary to revise Maxwell's laws and express the properties of electromagnetic systems by avoiding using the value of c as a constant: logically speaking this is irreproachable, and in fact this was the way tempted by a contemporary physicist of Einstein, Walter Ritz, who in 1908, aged thirty years, published the first voluminous attempts and results in this direction, and in 1909 died of tuberculosis.

3. Einstein's solution

The problem is that the revision of Maxwell's laws so as to avoid the use of the constant c, and therefore the reference to the ether, which Maxwell and the physicists of the nineteenth century did not question, is a very complex problem. Given this complexity, Einstein thought of looking for a much simpler path, which consists of the generalization of Lorentz's hypothesis. If we have realized that Lorentz's hypothesis solves the problem because the contraction of the systems in motion with respect to the ether would give rise to ever-equal measurements of the speed of light, we also understand, or rather, we make a first idea of how a generalization of the hypothesis could solve the problem even better, avoiding the artificial features of Lorentz's hypothesis. In any case, Lorentz's hypothesis helps us to give a preliminary, unclear meaning to the idea that the speed of light is always the same and cannot be added to any other speed.

Let us see exactly what happens in the conceptual framework of the ether without and with the hypothesis of Lorentz. If it is assumed that electromagnetic waves move in the ether, the speed of the body that emits them must never be added to the speed of light. Suppose that there is the ether, that a lamp at point A having precise x, y and z coordinates with respect to some origin in the ether, and that at a given instant the lamp emits a flash; when this happens, the flash propagates into the ether at speed c and the lamp can stay at point A or move away at any speed without affecting the flash speed. It does not matter whether the lamp is moving or is at rest with respect to the ether at the time of the flash shot. But in this hypothesis, measurements of the speed of light should reveal the Earth's speed with respect to the ether, also without the need to build a specific instrument like that of M+M. In fact, suppose for simplicity to be able to measure the journey of a flash going from point A to point B on the Earth surface, and suppose that the AB line is parallel to the direction of the Earth motion in the ether. If the flash is emitted in the sense of motion, the measurement of the speed of light will be smaller than that detected if the Earth were at rest compared to the Ether, because the B point will move away from the beam while it is travelling to reach it. If instead the flash is emitted in the opposite sense of the motion, then the measured speed would be greater than the one at rest, because point B would meet the beam. But this is never observed as a result of the contraction hypothesis. The first version of Lorentz's hypothesis assumed only spatial contraction, because this assumption was sufficient to explain the negative result of the M+M experiment, which does not include clocks or time measurements, but is designed to detect only the interference between two rays that travelled perpendicularly to each other. Then it was observed (by Larmor, as we have recalled) that hypothesizing only the contraction of lengths Lorentz's theory could not be explained at atomic level, given the theories of the electron of that time, and for this reason the contraction of durations was also introduced. But with or without the contraction of durations, within Lorentz's theory the measured speed of light will always be the same for all observers, regardless of their speed with respect to the ether, that is, of their speed with respect to the light rays. The shrinking of lengths alone, or of lengths and durations, cancels the effect of the observer's displacement in the ether and hence the measured speed of light is always c.

At this point we realize that Einstein's hypothesis has an aspect of analogy with the consequences of Lorentz's. Einstein's hypothesis, however, remains unacceptable without a further theory explaining how to accord it with the idea that light moves with respect to the emitting body and not to the ether. Meeting it for the first time, the reader could interpret it in a misleading way, as if it were saying: every device to measure the speed of light always gives the same result, both at rest and in motion. But the reader should not fall into this misunderstanding. If one speaks of a beam generated inside the apparatus, there is no difficulty in accepting the principle, and indeed, this is the empirical data we are in possession of: there are many variants of the device to measure the speed of light, each variant is based on the emission and reflection of a light beam inside the device, so there is a precise frame for the motion of the light whose speed is measured. Fizeau's device measured a beam that travelled 8 kilometres and a half in the way forth and the same in the way back, but the emitter was in Montmartre, and the mirror was on the roof of a country house outside Paris. The frame origin was fixed at the point of Paris where the beam whose journey was timed was emitted, and the same happens in all subsequent measurements.

3. Einstein's solution

But Einstein speaks of light beams emitted by a source that is at rest with respect to the apparatus, as well as light beams emitted by an external source with respect to which the apparatus is in motion, and postulates that the speed of light is always the same and that in both cases we must never add the speed of light to another one, i.e. to the speed of the source or that of the apparatus. What the postulate expresses is this: let us imagine we have an apparatus like that of Fizeau, which measures the speed of a beam generated by the apparatus, and let us imagine we have another apparatus capable of measuring with the same accuracy the speed of sun rays. The situation is hypothetical, because they did not exist in 1905 and still do not exist devices capable of measuring accurately the speed of light emitted by a source external to the apparatus. Let us suppose, however, that the Earth is approaching the Sun at 300 m/s, and that the Sun is behind us, close to sunset. Then, according to the new postulate, the apparatus measuring the beam emitted on the Earth and the apparatus measuring the beam emitted by the Sun would return the same value c. The apparatus that measures the beam emitted by the Sun would not return the value $c+300$ m/s, but just the value c. If one did the same operations on a train approaching the Sun at a speed of 100 m/s, he would still get the value c, and not $c+400$. This would happen in Lorentz's hypothesis, but readers may wonder: why formulate the strange hypothesis of extending this property to the case of absence of the ether?

In the ether hypothesis, the speed of light was considered relative to the ether, and therefore the measured speeds on the Earth would have to vary in relation to the Earth's motion of the Earth with respect to the ether at the time of measurement, but nothing happened. Then, abandoning the hypothesis of the ether, not wanting to adopt the artificial hypothesis of Lorentz to explain the observed phenomenon, and adopting the idea that light propagates in the vacuum, why simply not assume that light behaves like anything else given the law of inertia, and that its speed is always c with respect to the emitting body, but $c+v$ with respect to an observer who moves away or approaches the emitting body? This point of view was adopted by some, including Walter Ritz, whom we have already mentioned, and takes the name of *emission theory*; but it did not prevail, because there is a system of knowledge, electromagnetic theory, in which speed c is conceived and used as a constant, and therefore the change of perspective consisting in conceiving light as any other thing moving according to the law of inertia, would require to review and to correct the electromagnetic theory.

The postulate of the absolute constancy of c even in the absence of the ether is not easily understandable in any case, but it is at least conceivable, it has a meaning, if you know the hypothesis of Lorentz's contraction, that gives an idea of how a world could be structured in which the speed of light were constant in every case, and it seems motivated if one knows that there is a serious problem lying in the fact that speed c is used as an absolute constant by a fundamental branch of physics. Without these premise, the postulate seems both inconceivable and unmotivated. Knowing Lorentz's hypothesis, it is still unclear how the postulate could be conceivable if the ether that causes contraction is not assumed and if the law of inertia is generalized, but at least one starts with an idea of what could be the meaning of words and how Einstein's hypothesis could be effective in the physical world, and above all one starts with a clear notion of the need to which Einstein's hypothesis tries to find a solution. It was necessary to preserve the form of the laws of Maxwell, and Lorentz's way was that of the contraction of spatial lengths and time duration. Einstein proposed to adapt all this to the hypothesis that the ether does not exist and that electromagnetic waves propagate simply in vacuum moving with respect to the emitting body and not with respect to the ether or any other absolute frame.

Here is an observation on relativistic literature: almost all the textbooks of every level just hint at the hypothesis of Lorentz's contraction without fully describing it, and then proceed to develop the application of Lorentz's mathematics directly from Einstein's viewpoint. Moreover, concentrated on the description of Einstein's theory, relativistic literature either keeps completely silent about the electrodynamic issue, or leaves it in the background: instead, this issue must be known and understood, because it is the heart of the matter, and provides the motivation of Einstein's hypothesis. The paper of 1905 has the title "On the electrodynamics of moving bodies" because this is the central problem, not by chance.

3. Einstein's solution

From the point of view of expressive effectiveness, the result of relativistic accounts is disastrous: readers are assaulted with the notion of the constancy of the speed of light, and if they understand what they read, they remain with the impression of an inconceivable and absurd postulate. It remains in their mind that the speed of light should, in their view, be added to that of the source, and therefore be relative to the frame chosen to measure it, as in our example of the motorbike, and consequently they either abandon the study of the theory of special relativity thinking humbly of not being able to understand it, or assume the postulate of the constancy of speed of light as a *flatus vocis*, an expression that does not correspond to any definite meaning, to be assimilated as a kind of article of faith. Therefore, the knowledge of both the historical precedent of Lorentz's hypothesis and of the electrodynamic problem is indispensable to have an actual and non-fictive understanding of the generalization attempted by Einstein.

3.2   Synchronization of clocks in distant systems

At this point, Einstein has to prepare the ground to apply Lorentz's math to any uniform rectilinear motion, even without the absolute frame of the ether. It is therefore necessary to find a kinematics in which this is conceivable. Due to Lorentz's mathematics, length and duration measurements, which classical mechanics assume stable, are subject to change as a consequence of speed with respect to the ether; Einstein wants to find the key to apply the same variations of measure simply as consequences of motion and speed with respect to anything, and for this purpose in ED1905, in chapters 1 and 2, he describes a situation in which systems moving with respect to each other at constant speed would find discordant measures of time and space as a consequence of having adopted a certain synchronization procedure for their clocks. The same situation is described by Einstein in chapters 3 to 10 of his popular account[18] of the theory, with a considerable number of pages, but to grasp the question it is preferable to develop the reasoning without the digressions of the simplified account, following the sequence of the original text and integrating it with considerations absent in the succinct expression of Einstein.

Einstein's biographies inform us that Einstein, at that time 26 years old and employed at the Swiss Federal Bureau of Patents, was stimulated to reflect on the principle of synchronizing clocks that are distant from each other by some of the inventions that required a patent and were therefore presented to his office for technical examination, and by the daily show of the railway station in Bern, which used the clock of the old city tower as the main reference to determine the standard time of the Swiss railways[19]. These extrinsic factors are not relevant to the theoretical issue, but place the idea in a precise technological context: that of the first steps of automatic transmissions through electromagnetic signals. Meanwhile Marconi was building his first radio stations on the Irish coast to assist the Atlantic navigation.

---

[18] Einstein 1916.

[19] See Isaacson 2007, Chapter 6 and previous chapters.

The situation imagined by Einstein is this.

There are two so-called "observers", two people A and B, who conclude a convention to synchronize their clocks while they are distant from each other and cannot match the clocks with a simple direct observation of both the dials, because they are not in the same place.

First, Einstein assumes that clocks exist which can mark the same time, and that A and B are provided with them. B possesses "eine Uhr von genau derselben Beschaffenheit wie die in A befindliche", i.e. "a clock that has exactly the same characteristics as that in A", and there is no further discussion of the clock concept. In a much more recent text we find this description[20]:

> The primitive subjective feeling of time flow enables us to order our impressions, to judge that one event takes place earlier, another later. But to show that the time interval between two events is to seconds, a clock is needed. By the use of a clock the time concept becomes objective. Any physical phenomenon may be used as a clock, provided it can be exactly repeated as many times as desired. Taking the interval between the beginning and the end of such an event as one unit of time, arbitrary time intervals may be measured by repetition of this physical process. All clocks, from the simple hour-glass to the most refined instruments, are based on this idea.

The concept of clock is all here: a counter of the repetitions of an event that we have assumed to repeat themselves being always identical, without ever knowing if there is an absolute clock worthy of being privileged to each other.

Observers A and B have two clocks that, placed next to each other, having "the same characteristics" would mark the same time. They want to tune the measurements of the two clocks by setting them to zero at the same time instant. To synchronize clocks while the two observers are distant, they agree a procedure that is described in a simple way, as if it were carried out by people with their stopwatches in hand, but which obviously should be understood as a schematic illustration of an ideal procedure. One must imagine, as an immediate next step, that the procedure could be performed by automated electrical devices designed specifically for the purpose,

---

[20] Einstein and Infeld 1938, Chapter "Time, Distance, Relativity".

but this would still serve only to help understanding with the example. As a further step we must understand that for the argument only the logical sequence of events is important, and that the procedure is completely ideal. However, it is important to note that, even if imagined as executed by automated equipment, even if properly conceived as purely ideal, it is assumed that the procedure would be designed and interpreted by thinking minds, who have the necessity and the will to carry out space and time measurements: that is, the procedure does not describe the operation schema of a natural phenomenon, but of a human act.

The procedure imagined by Einstein assumes the use of either an electric signal, or a light or a radio signal. In any case, a signal that travels at speeds c, which has been assumed constant in the premises of the theory. The steps of the procedure are:

- A resets its clock and sends an electrical or light or radio signal to B.

- B receives the signal and reflects it towards A, and exactly at that moment it resets its clock; if we think of an automatic procedure, we can imagine that a device is built in B that receives the signal and sends it back to A in a tiny, negligible time, and while performing this action it resets its clock.

- A receives the return signal, reads the time t marked by its clock, and sets its clock to $t/2$.

If A and B had performed this procedure at a certain distance, but at rest with respect to each other, the two clocks would certainly take the same time because the signal would travel from A to B in time $t/2$ and would return from B to A again in $t/2$. Hence, in the case of relative rest, when A reads the time and corrects its clock setting it back to $t/2$, the clock B also marks $t/2$ and since the two clocks "have the same characteristics", from now on they will mark the same time, until one of the two pauses or fails.

The argument is very simple, but let us also develop a numerical example. A sets its clock on zero and sends the signal. After 5 seconds the signal reaches B and is reflected. B sets its clock on zero. After another 5 seconds the signal comes to A. A reads 10 seconds (5+5) and corrects its clock on the value of 5 seconds. Since 5 seconds have elapsed since B has reset its clock, the B clock also marks 5 seconds. The situation is:

Event	The clock in A marks	The clock in B marks
The signal starts from A	0	nothing
The signal arrives at B	5	0
The signal comes back to A	5+5=10	5
A corrects its clock	10/2=5	5

Now we need to see what happens if this procedure is performed involving observers in motion. Einstein, after defining the synchronization procedure in this way, develops the case of observers in motion with seemingly simple, but very abstract considerations. To fully understand Einstein's theory, we have to examine carefully two elementary examples of a different nature.

3.3   Synchronization of distant clocks with other means than light

Let us imagine there are two wagons running slowly in the country at constant speed v for both, in the same direction and in the same sense. Suppose they are at a distance of a few kilometres from each other, with A that follows B, and that they agreed to synchronize by resetting their clocks with a procedure similar to that described above. They have two ways to proceed: either by sending a horseman travelling faster than the two wagons, or by shooting an arrow to each other. Of course, both synchronization means have to travel at greater speed than v, because otherwise the two wagons could not communicate by signal exchange.

First case: synchronization with a horseman.

Wagon A resets its clock and sends the horseman with the order to tell B "reset your clock soon" and come back immediately. The horseman travels at constant speed w; w is therefore the speed of the signal. A and B are at rest with respect to each other because the speed with which they travel is v for both and the direction is the same, but the horseman's speed w is not in the frame of A and B, but is measured with respect to the road on which the horseman moves. So if A and B assumed that their clocks would mark the same time as they have adopted the synchronization procedure we know, they would be wrong, because on the way forth the carriage B escapes the horseman at speed v with respect to the ground while returning the carriage A meets the horseman at speed v. In the journey forth the horseman takes a time t', returning a time t'', which is less than t', because the horseman's speed with respect to the ground is always the same, but on the way forth the carriage B escapes him, and at return the carriage A meets him.

Let us see a numerical example. We set speed of the carriages$=v=5$ m/s, speed of the horseman (signal)$=w=20$m/s, constant distance between carriages$=AB=45$ m.

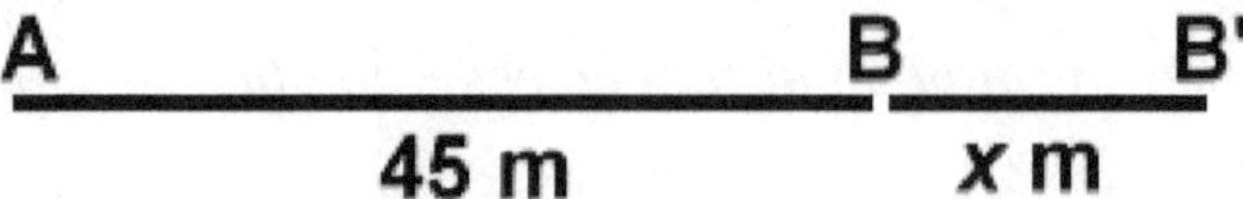

The horseman leaves from A and must reach B. In the meantime, the carriage B moves to B'. So while the horseman runs the AB' segment, or AB+BB', the carriage runs BB'. We have to find the time t' of the way forward.

We have:

$$\frac{45+BB'}{t'} = 20 = \text{horseman's speed;}$$

$$\frac{BB'}{t'} = 5 = \text{speed of wagon B.}$$

Hence:

$$t' = \frac{BB'}{5}$$

and replacing:

$$\frac{45+BB'}{\dfrac{BB'}{5}} = 20\,;$$

$$\frac{5(45 + BB')}{BB'} = 20\,;$$

$$45 + BB' = 4(BB');$$

$$45 = 3(BB');$$

$$BB' = 15\,.$$

Time t' is:

$$t' = \frac{15}{5} = 3.$$

Hence the horseman on the way forth runs 45+15=60 metres in the time of 3 seconds.

On the way back, the horseman leaves from B', and reaches carriage A which was in A', but which in the meantime meets the horseman and reaches point A".

This time the horseman runs the B'A" way while the carriage A runs through A'A". When the horseman reverses the sense, he is in B' and the carriage A is in A', and the A'B' segment is always 45 metres long, because the carriages proceed at the same speed and thus always keep the same distance.

We have:

$$\frac{45 - A'A''}{t''} = 20 = \text{horseman's speed};$$

$$\frac{A'A''}{t''} = 5 = \text{speed of wagon A}.$$

Hence:

$$t'' = \frac{A'A''}{5}$$

and replacing:

$$\frac{45 - A'A''}{\frac{A'A''}{5}} = 20 \, ;$$

$$\frac{5(45 - A'A'')}{A'A''} = 20 \, ;$$

$$45 - A'A'' = 4(A'A'') \, ;$$

$$45 = 5(A'A'') \, ;$$

$$A'A'' = 9 \, ;$$

Time t'' is:

$$t'' = \frac{9}{5} = 1.8 \, .$$

So we have the following situation:

Event	The clock in A marks	The clock in B marks
The horseman starts from A	0	nothing
The horseman arrives at B	3	0
The horseman comes back to A	3+1.8=4.8	1.8
A corrects its clock	4.8/2=2.4	1.8

3. Einstein's solution

In the end, A divides the overall time by half, and finds that its clock marks 2.4 seconds and believes that the clock B also marks 2.4 seconds, while the clock B marks only 1.8 seconds. The clock B is therefore late compared to A's clock for merely a logical reason, linked to the technique used for synchronization.

Einstein does not develop a reasoning equivalent to this example (it would be trivial), but similar to this example. The example is useful to get into the idea. We begin to see something that resembles Lorentz's contraction, but that has nothing to do with a physical slowdown of the processes that take place in motion with respect to the ether. There is nothing like that, and the two clocks disagree as the two observers have applied a certain procedure, and precisely being in motion they have applied (not legitimately) the same procedure that if applied in a state of rest with respect to the Earth would have synchronized the clocks exactly.

Let us also note that the example of the horseman's synchronization describes the problem that would arise if the ether existed. If we wanted to perform the synchronization procedure by means of a signal, we would be in trouble: two distant systems A and B at rest with respect to each other should know their speed with respect to the ether, because otherwise the synchronization procedure would result in the same mistake made by the carriage drivers of our example, who, ignoring the fact that they were in motion with respect to the Earth, found themselves to have discordant clocks. But the Earth's speed with respect to the ether is unknown, so synchronization through electromagnetic signals would give tiny, but theoretically insoluble, undetectable errors.

Second case: synchronization with an arrow.

Instead of sending a messenger on horse, A sends an arrow to B, and B resets the clock and replies by shooting the arrow back. While in the case of a horseman we could assume that he travelled at constant speed with respect to the ground, in the case of the arrow we cannot consider the speed constant because the arrow reaches its maximum speed while the bow rope accelerates it and then decelerates for the rest of its journey. However, given the symmetry of the situation, we can overlook deceleration, which will be the same on the AB and BA travel, and consider the average speed of the arrow, w. The speed (initial and average) of the arrow, however, is not relative to the ground, as was the speed of the horseman: the speed of the arrow must be added to that of the archer. As we have seen in section 2.4, this is always true in the ballistic case: if one shoots a gun in the sense of the motion in a train running at 100 m/s, and if the bullet speed when it comes out of the gun barrel is 900 m/s, then the bullet speed is 900 m/s with respect to the train, but 1000 m/s to the ground. Considering this, assume that the arrow starts at speed w with respect to the archer. On the AB route the arrow is shot in the sense of motion, so the speed of the archer must be added, and the arrow starts at the speed $v+w$ with respect to the ground. However, the carriage B moves away at speed v, so the speed v of the carriage B must be subtracted, hence the arrow moves towards B at the overall speed $v+w-v=w$.

On the return journey the opposite happens. The archer on carriage B shoots the arrow in the opposite sense to the motion, and then the arrow starts at speed $-v+w$ over the ground. But A meets, so speed v of A must be added, and then the arrow moves to A at the speed $-v+w+v=w$. Therefore by using the arrow method, the synchronization procedure leads the clocks to mark the same time in each case, whether the wagons A and B are stationary or are in motion with respect to the Earth. Indeed, the two wagons are at relative rest with respect to each other, and the state of motion over the Earth has no influence on the procedure.

3.4   *Lorentz's transformations come into play*

Let us now think according to Einstein's premises. We have abandoned the ether hypothesis, both because the experiments have refused to reveal the Earth's speed with respect to ether, and because some fundamental method requirements want us to apply the law of inertia to any field of physics.

Then, we should move on to conceive the motion of light similar to that of bullets: light starts at the speed c from the emitter and this sums algebraically to the speed of the emitting body. The speed of light becomes relative as any other speed is. So, for example, if at some instant the Earth moves towards the Sun approaching it at 300 m/s and Mars moves away from the Sun at a speed of 500 m/s (remember that planets describe an elliptical orbit around the Sun, which is in one of the foci of the ellipse, therefore in addition to revolving around the Sun planets are also always approaching or moving away from it), then the speed of the sun rays would be c+300 m/s for the inhabitants of the Earth and c-500 m/s for the inhabitants of Mars.

You might ask: what is strange about this? If there is no ether, and if we do not want to admit absolute frames for the motion of any physical reality, it will be like this.

But Einstein in 1905 would object that adopting this model instead of that of the ether, would require to reconsider the whole electromagnetic theory, and to modify the form of Maxwell's equations, because in them the speed of light appears as a constant.

So Einstein says, and this is the heart of the entirety of special relativity: let us hypothesize all the elements at stake, and assume that in the example just made the Earth's dwellers and those of Mars, measuring the speed of light emitted by the Sun, find c, always c, and not c+300 m/s on the Earth or c-500 m/s on Mars.

In the foreword of Einstein's paper, he had stated the purpose of this hypothesis. Let us read again[21]:

> ... and shall introduce, in addition, the postulate, only seemingly incompatible with the former one, that in empty space light is always

---

[21] ED1905, Foreword.

propagated with a definite speed c which is independent of the state of motion of the emitting body.

Now, we know that in Lorentz's hypothesis all those who measure the speed of light would find the value c, because the spatial segments of the equipment to measure it would be contracted by the Lorentz gamma factor and the duration of the time segments would do the same, that is, the clocks used would slow down.

Therefore Einstein says: what must be done to ensure that the speed of light is always c? We just must apply the gamma factor and Lorentz transformations to all the inertial systems in relative motion: the speed of light remains constant, and Maxwell's equations consequently do not require revision. It follows that the simultaneity of events becomes relative: in fact, as we know, by applying the transformations of Lorentz, the times t and t' of the same event seen in two systems K and K' do not have the same numeric value. And it follows that the inertial systems K and K' undergo something like a mutual contraction, which we can call apparent, and which concerns only their measures taken by each other.

3.5   A question to the reader, and instructions

Do you think we are on the road to good reasoning, or is there a difficulty somewhere? In either case, whether the reasoning seems indisputable, or whether you feel something wrong, continue reading. Just make sure you have grasped the literal meaning of the things said so far, without hastening to judge Einstein's theory. Only if you have doubts about the meaning of what has been said in this book until now, not on the plausibility of the theory that is taking shape, go back and read again.

If you have doubts about the plausibility and consistency of Einstein's theory that we are describing step by step, continue reading: it is not yet time to consider such doubts.

3.6   *Synchronization of distant clocks by means of light*

Einstein, when interviewed many years later, sometimes said that when he wrote ED1905 he was unaware of Lorentz's hypothesis and transformations. This was called into question, and it was argued that Einstein certainly had a more or less profound knowledge of the hypothesis at least through Poincaré, and that perhaps many years later he did not remember exactly what he had read before 1905[22]. However, chapters 1 and 2 of ED1905 are structured so as to highlight the difference between the new theory and Lorentz's theory, with which it shares the formulas. In these chapters, Einstein describes a situation that serves to prepare the ground for Lorentz's transformations, writing a hard text to read and not even exactly accurate, when he could quote Lorentz's hypothesis and exploit the work already done, as we did. Einstein, knowing Lorentz's theory, could say: Lorentz's theory hypothesizes ether and physical contraction, but this is not the only case in which Lorentz's mathematical formulas can be applied to two inertial systems. They may, under certain conditions, apply to reciprocally moving systems without even hypothesizing the contraction in the direction of ether wind. And that said he could develop his idea. Instead, Einstein chose to start from scratch and to create a situation to which Lorentz's formulas could apply through a description that is independent of the antecedents.

In Chapter 1, whose title is "Definition of simultaneity", after some principle assertions around the concept of time, Einstein presents the method we described for synchronizing two remote clocks. He assumes that there is no ether and there is no problem of speed with respect to the ether, that is, the problem of our example of the two wagons and the horseman, so that between two clocks A and B the procedure is always correct, provided that A and B are each at rest with respect to the other, without the fact that they are in motion or at rest with respect to any other frame has any relevance. Things take place as in our example of the arrow.

---

[22] See Trefil 2015 Chapter 4 and Isaacson 2007 passim.

That said, he also observes that synchronization can take place in a chain of clocks, and if A is synchronized with B and B with C, then A is also synchronized with C, and so on.

Chapter 2 instead deals with clock synchronization within a situation when there are two systems K and K' moving one with respect of each other, imagining as always for simplicity that motion takes place along the x-axis. Each of the two systems K and K' knows the synchronization events in the other system and gives judgments about them. The two chapters each contain a few paragraphs, which we shall read wholly and comment on.

3.7   *ED1905 Chapter 1, "Definition of simultaneity"*

The chapter has the title "§1. Definition der Gleichzeitigkeit", i.e. "Definition of simultaneity", and contains ten paragraphs, which we shall now comment on one by one. There are a few footnotes, here below reproduced after the paragraphs to which they refer.

§1, Paragraph 1.

Es liege ein Koordinatensystem vor, in welchem die Newtonschen mechanischen Gleichungen gelten. Wir nennen dies Koordinatensystem zur sprachlichen Unterscheidung von später einzuführenden Koordinatensystemen und zur Präzisierung der Vorstellung das "ruhende System".

Consider a coordinate system in which the Newtonian mechanical equations are valid. To distinguish it verbally from the coordinate systems that will be introduced later on, and to visualize it more precisely, we shall designate this system as the "system at rest."

No difficulty in interpretation. The concepts of "system at rest" and "system in motion" will be used, knowing that the attribution of the rest and motion condition will always be arbitrary.

§1, Paragraph 2.

Ruht ein materieller Punkt relativ zu diesem Koordinatensystem, so kann seine Lage relativ zu letzterem durch starre Maßstäbe unter Benutzung der Methoden der euklidischen Geometrie bestimmt und in kartesischen Koordinaten ausgedrückt werden.

If a material point is at rest relative to this coordinate system, its position relative to the latter can be determined by means of rigid measuring rods using the methods of Euclidean geometry and can be expressed in Cartesian coordinates.

The fundamental measure of space in the system at rest is conceived as easily as possible: a length, the distance between two points on an axis, is the multiple of the length of a rigid ruler assumed as a unit of measure. Earlier it was assumed that there are clocks that can indicate the same time, it is now assumed that there are rulers that indicate lengths that are stable at rest. Nothing is said about the conventions needed to define the length sample.

§1, Paragraph 3.

Wollen wir die Bewegung *eines materiellen Punktes beschreiben, so geben wir die Werte seiner Koordinaten in Funktion der Zeit. Es ist nun wohl im Auge zu behalten, daß eine derartige mathematische Beschreibung erst dann einen physikalischen Sinn hat, wenn man sich vorher darüber klar geworden ist, was hier unter "Zeit" verstanden wird. Wir haben zu berücksichtigen, daß alle unsere Urteile, in welchen die Zeit eine Rolle spielt, immer Urteile über* gleichzeitige Ereignisse *sind. Wenn ich z. B. sage: "Jener Zug kommt hier um 7 Uhr an," so heißt dies etwa: "Das Zeigen des kleinen Zeigers meiner Uhr auf 7 und das Ankommen des Zuges sind gleichzeitige Ereignisse."*

Anmerkung: Die Ungenauigkeit, welche in dem Begriffe der Gleichzeitigkeit zweier Ereignisse an (annähernd) demselben Orte steckt und gleichfalls durch eine Abstraktion überbrückt werden muß, soll hier nicht erörtert werden.

If we want to describe the *motion* of a material point, we give the values of its coordinates as a function of time. However, we should keep in mind that for such a mathematical description to have physical meaning, we first have to clarify what is to be understood here by "time." We have to bear in mind that all our propositions involving time are always propositions about *simultaneous events*. If, for example, I say that "the train arrives here at 7 o'clock," that means, more or less, "the pointing of the small hand of my clock to 7 and the arrival of the train are simultaneous events."

Footnote: We shall not discuss here the imprecision that is inherent in the concept of simultaneity of two events taking place at (approximately) the same location and that also must be surmounted by an abstraction.

To describe the motion of a point, we must give the values of its space coordinates as a function of time, i.e. we must describe how space coordinates change in time. To measure space, we have defined an elemental operation, that of superimposing a ruler to the object to be measured. To measure time, we define the elemental operation that consists of looking at the phenomenon to be determined and at the clock by focusing on both. The footnote states that obviously the eye measurements made in this way include errors due to the perceptional human apparatus, and that we must be able to suppress such errors by abstraction. We have to interpret it with some elasticity: for example, the act of watching a runner and stopping the stopwatch with one's thumb while the runner cuts the finish line (without the need to look at the stopwatch) is conceptually equivalent to the method of determining a time watching the event and the clock together. The meaning of the discourse is that the fundamental technique for measuring time is to notice the simultaneity between events and duration samples, or between events in the object to be measured and events in the clock, and that approximations due to the psychological characteristics of the human brain are not in question, because they are overlooked by the hypothesis.

§1, Paragraph 4.

Es könnte scheinen, daß alle die Definition der "Zeit" betreffenden Schwierigkeiten dadurch überwunden werden könnten, daß ich an Stelle der "Zeit" die "Stellung des kleinen Zeigers meiner Uhr" setze. Eine solche Definition genügt in der Tat, wenn es sich darum handelt, eine Zeit zu definieren ausschließlich für den Ort, an welchem sich die Uhr eben befindet; die Definition genügt aber nicht mehr, sobald es sich darum handelt, an verschiedenen Orten stattfindende Ereignisreihen miteinander zeitlich zu verknüpfen, oder — was auf dasselbe hinausläuft — Ereignisse zeitlich zu werten, welche in von der Uhr entfernten Orten stattfinden.

It might seem that all difficulties involved in the definition of "time" could be overcome by my substituting "position of the small hand of my clock" for "time." Such a definition is indeed sufficient if time has to be defined exclusively for the place at which the clock is located; but the definition becomes insufficient as soon as series of events occurring at different locations have to be linked temporally, or — what amounts to the same — events occurring at places remote from the clock have to be evaluated temporally.

It is noted that the method of watching together the event and the clock is practicable only in a space close to the observer. Determining the time of an event occurring inside a plane flying over us obviously cannot be done with such a simple technique, and one can find countless elementary examples of this.

§1, Paragraph 5.

Wir könnten uns allerdings damit begnügen, die Ereignisse dadurch zeitlich zu werten, daß ein samt der Uhr im Koordinatenursprung befindlicher Beobachter jedem von einem zu wertenden Ereignis Zeugnis gebenden, durch den leeren Raum zu ihm gelangenden Lichtzeichen die entsprechende Uhrzeigerstellung zuordnet. Eine solche Zuordnung bringt aber den Übelstand mit sich, daß sie vom Standpunkte des mit der Uhr versehenen Beobachters nicht unabhängig ist, wie wir durch die Erfahrung wissen. Zu einer weit praktischeren Festsetzung gelangen wir durch folgende Betrachtung.

To be sure, we could content ourselves with evaluating the time of the events by stationing an observer with the clock at the coordinate origin, and having him assign the corresponding clock-hand position to each light signal that attests to an event to be evaluated and reaches him through empty space. But as we know from experience, such an assignment has the drawback that it is not independent of the position of the observer equipped with the clock. We arrive at a far more practical arrangement by the following consideration.

The meaning of this paragraph bears some difficulty, especially for the words "as we know from experience". In the previous section it was said that there are situations where time measurement to the eye cannot be done because events to be determined temporally are not observable in a tight local environment. Here it is said that there is another problem, connected with the case where a distant observer measures with the eye the time of remote events he or she can see. This observer should also take into account the time taken by the signal to reach him or her, and if he or she did not know the distance between him or herself and the event, he or she would be in trouble. But why is "experience" invoked? And why is there the problem of finding a "more practical" method? Probably the answer is that Einstein was influenced to think of this by observing the synchronization methods used by Swiss railways ("experience"), and the patent proposals for electrical synchronization mechanisms that he had to evaluate in the office where he was employed ("more practical methods"). The experience mentioned is that of synchronization of the train stations clocks, which was difficult with the technologies of those times, and the practical need was to solve the organizational problems of the railways. A biographer says:[23] "Among the ideas that he had to consider for patents were dozens of new methods for synchronizing clocks and coordinating time through signals sent at the speed of light."

But the text here is inappropriate and misleading, because there was no need to make this hint to an empirical perspective without explaining its exact meaning; and the technological problem (important in 1905 and negligible today, in the era of electronics) is irrelevant to the theory that is being built, either at that time or today. What matters is to define the remote synchronization procedure in an unambiguous way, which occurs in the following paragraph:

§1, Paragraph 6.

Befindet sich im Punkte A des Raumes eine Uhr, so kann ein in A befindlicher Beobachter die Ereignisse in der unmittelbaren Umgebung von A zeitlich werten durch Aufsuchen der mit diesen Ereignissen gleichzeitigen Uhr Zeigerstellungen. Befindet sich auch im Punkte B des Raumes eine Uhr — wir wollen hinzufügen, "eine Uhr von genau derselben Beschaffenheit wie die in A befindliche" — so ist auch eine zeitliche Wertung der Ereignisse in der unmittelbaren Umgebung von B durch einen in B befindlichen Beobachter möglich. Es ist aber ohne weitere Festsetzung nicht möglich, ein Ereignis in A mit einem Ereignis in B zeitlich zu vergleichen; wir haben bisher nur eine "A-Zeit" und eine "B-Zeit", aber keine für A und B gemeinsame "Zeit" definiert. Die letztere Zeit kann nun definiert werden, indem man durch Definition festsetzt, daß die "Zeit", welche das Licht braucht, um von A nach B zu gelangen, gleich ist der "Zeit", welche es braucht, um von B nach A zu gelangen. Es gehe nämlich ein Lichtstrahl zur "A-Zeit" t_A von A nach B ab, werde zur "B-Zeit" t_B in B gegen A zu reflektiert und gelange zur "A-Zeit" t'_A nach A zurück. Die beiden Uhren laufen definitionsgemäß synchron, wenn

$$t_B - t_A = t'_A - t_B.$$

If there is a clock at point A of space, then an observer located at A can evaluate the time of the events in the immediate vicinity of A by finding the clock-hand positions that are simultaneous with these events. If there is also a clock at point B — we should add, "a clock of exactly the same

---

[23] See Isaacson 2007, Chapter 4.

constitution as that at A" − then the time of the events in the immediate vicinity of B can likewise be evaluated by an observer located at B. But it is not possible to compare the time of an event at A with one at B without a further stipulation; thus far we have only defined an "A-time" and a "B-time" but not a "time" common to A and B. The latter can now be determined by establishing *by definition* that the "time" needed for the light to travel from A to B is equal to the "time" it needs to travel from B to A. For, suppose a beam of light leaves from A toward B at "A-time" t_A, is reflected from B toward A at "B-time" t_B, and arrives back at A at "A-time" t'_A. The two clocks are synchronous by definition if

$$t_B - t_A = t'_A - t_B.$$

Here begins the description of the synchronization mechanism, which is what we already know. To have exact clock synchronization with a light beam, we must assume that the forward time and the return time of a beam are equal. If, to carry the signal, instead of the light we used a horseman, as in the previous example, even if A and B were at rest with respect to each other, we could never have the guarantee that the signal travels at the same speed on the way forth and on the return: the rider could travel at different speeds for infinite reasons. Using light, we may also have doubts: in the hypothesis of the ether, we should take into account the speed and the sense of the Earth's motion with respect to it to correct the duration of the forward and return journey. But even without the hypothesis of the ether, we cannot know that the speed of light does not absolutely change according to some condition. So the constancy of the speed of light is postulated, the assertion that the return journey takes the same time as the one forward is also a postulated assumption, and therefore the synchronization method based on it is a convention. Beware, however: here is postulated that the light has constant speed only in a generic sense, which would be valid also in the case in which c is to be added to the speed of the source, not in the specific sense of Einstein's second postulate. The meaning of the argument in this paragraph is that as we always choose the sample clock without being able to know whether it is absolutely regular compared to an absolute clock, so to synchronize clocks remotely we assume that the light propagates at constant speed without knowing whether it is constant with respect to some absolute evaluation criteria.

§1, Paragraph 7.

Wir nehmen an, daß diese Definition des Synchronismus in widerspruchsfreier Weise möglich sei, und zwar für beliebig viele Punkte, daß also allgemein die Beziehungen gelten:

1. Wenn die Uhr in B synchron mit der Uhr in A läuft, so läuft die Uhr in A synchron mit der Uhr in B.

2. Wenn die Uhr in A sowohl mit der Uhr in B als auch mit der Uhr in C synchron läuft, so laufen auch die Uhren in B und C synchron relativ zueinander.

We assume that it is possible for this definition of synchronism to be free of contradictions, and to be so for arbitrarily many points, and that the following relations are therefore generally valid:

1. If the clock in B is synchronous with the clock in A, then the clock in A is synchronous with the clock in B.

2. If the clock in A is synchronous with the clock in B as well as with the clock in C, then the clocks in B and C are also synchronous relative to each other.

Simple result to infer. The synchronization is mutually valid for two clocks, and it is possible to synchronize many concatenated clocks.

§1, Paragraph 8.

Wir haben so unter Zuhilfenahme gewisser (gedachter) physikalischer Erfahrungen festgelegt, was unter synchron laufenden, an verschiedenen Orten befindlichen, ruhenden Uhren zu verstehen ist und damit offenbar eine Definition von "gleichzeitig" und "Zeit" gewonnen. Die "Zeit" eines Ereignisses ist die mit dem Ereignis gleichzeitige Angabe einer am Orte des Ereignisses befindlichen, ruhenden Uhr, welche mit einer bestimmten, ruhenden Uhr, und zwar für alle Zeitbestimmungen mit der nämlichen Uhr, synchron läuft.

With the help of some physical (thought) experiments, we have thus laid down what is to be understood by synchronous clocks at rest that are situated at different places, and have obviously obtained thereby a definition of "synchronous" and of "time." The "time" of an event is the reading obtained simultaneously with the event from a clock at rest that is located at the place of the event and that for all time determinations is in synchrony with a specified clock at rest.

Conclusion without interpretative problems, except for a little convoluted language.

§1, Paragraph 9.

Wir setzen noch der Erfahrung gemäß fest, daß die Größe

$$\frac{2AB}{t'_A - t_A} = c$$

eine universelle Konstante (die Lichtgeschwindigkeit im leeren Raume) sei.

Based on experience, we also postulate that the quantity

$$\frac{2AB}{t'_A - t_A} = c$$

is a universal constant (the speed of light in empty space).

Since AB is the distance between the two clocks and t'_A-t_A is the total time of the return trip of the beam, twice the distance divided by the total time will give us the speed of light. And this is a universal constant, or independent of the speed of the source: now the convention corresponds fully to the second postulate. Experience is also mentioned here, and we know what experiences it is about: the many attempts to determine the Earth's speed with respect to the ether, which have always given the same null result, and the fact that the apparatus to measure the speed of light have always returned the same value as a result of the failure of determining the Earth's speed with respect to the ether.

However, one thing has to be clarified: all measurements have always been made by means of devices that measure a beam emitted within the equipment and reflected within the equipment. All the equipment are evolutions of that of Fizeau and Michelson and Morley, which we have described. So no experience has shown (either at that time or today) that, for example, the speed of a Sun beam would not be equal to $c+v$, if v is the speed of the Earth approaching to the Sun, and to $c-v$ if the Earth is moving away. The absolute constancy of c is a postulate, with no direct empirical evidence, because experience has shown only 1) that one cannot determine the Earth's speed with respect to the ether, and 2) that the measure of beams generated by an apparatus and reflected within it always gives the same result, which is consistent with the emission theory and does not require Einstein's second postulate. No experience has ever shown that the speed of light should not be added to that of the measurement equipment, and therefore consistently with this, Einstein insists that the absolute constancy of c is an assumption that does not conflict with empirical data ("der Erfahrung gemäß"), but which could theoretically be replaced by the assumption that the speed of light must be added to that of the source. Emission theory, according to which light propagates in the vacuum and according to which the constant c is simply the speed of light with respect to the emitting body, is excluded by assumption, not because experience proves it impossible.

§1, Paragraph 10.

Wesentlich ist, daß wir die Zeit mittels im ruhenden System ruhender Uhren definiert haben; wir nennen die eben definierte Zeit wegen dieser Zugehörigkeit zum ruhenden System "die Zeit des ruhenden Systems".

It is essential that we have defined time by means of clocks at rest in a system at rest; because it belongs to the system at rest, we designate the time just defined as "the time of the system at rest."

These final words point out the unambiguous character of the given definitions.

3.8   ED1905 Chapter 2, "On the relativity of lengths and times"

The second chapter has the title "Über die Relativität von Längen und Zeiten", "On the relativity of lengths and times", and consists of the following eight paragraphs:

§2, Paragraph 1.

Die folgenden Überlegungen stützen sich auf das Relativitätsprinzip und auf das Prinzip der Konstanz der Lichtgeschwindigkeit, welche beiden Prinzipien wir folgendermaßen definieren.

1. Die Gesetze, nach denen sich die Zustände der physikalischen Systeme ändern, sind unabhängig davon, auf welches von zwei relativ zueinander in gleichförmiger Translationsbewegung befindlichen Koordinatensystemen diese Zustandsänderungen bezogen werden.

2. Jeder Lichtstrahl bewegt sich im "ruhenden" Koordinatensystem mit der bestimmten Geschwindigkeit c, unabhängig davon, ob dieser Lichtstrahl von einem ruhenden oder bewegten Körper emittiert ist. Hierbei ist

Geschwindigkeit=Lichtweg/Zeitdauer,

wobei "Zeitdauer" im Sinne der Definition des §1 aufzufassen ist.

The considerations that follow are based on the principle of relativity and the principle of the constancy of the speed of light, two principles that we define as follows:

1. The laws governing the changes of the state of any physical system do not depend on which one of two coordinate systems in uniform translational motion relative to each other these changes of the state are referred to.

2. Each beam of light moves in the coordinate system "at rest" with the definite speed c independent of whether this beam of light is emitted by a body at rest or a body in motion. Here,

speed=light path/time interval,

where "time interval" should be understood in the sense of the definition in §1.

The principles set out in the foreword are expressed again, with the clarification that the concept of time interval has now been given a precise definition: one determines a time value either by looking at the event and the clock together, or by synchronizing two clocks by the procedure in which a light beam is sent by the near clock and returned by the remote clock.

§2, Paragraph 2.

Es sei ein ruhender starrer Stab gegeben; derselbe besitze, mit einem ebenfalls ruhenden Maßstabe gemessen, die Länge l. Wir denken uns nun die Stabachse in die X-Achse des ruhenden Koordinatensystems gelegt und dem Stabe hierauf eine gleichförmige Paralleltranslationsbewegung (Geschwindigkeit v) längs der X-Achse im Sinne der wachsenden x erteilt. Wir fragen nun nach der Länge des bewegten Stabes, welche wir uns durch folgende zwei Operationen ermittelt denken:

a) Der Beobachter bewegt sich samt dem vorher genannten Maßstabe mit dem auszumessenden Stabe und mißt direkt durch Anlegen des Maßstabes die Länge des Stabes, ebenso, wie wenn sich auszumessender Stab, Beobachter und Maßstab in Ruhe befänden.

b) Der Beobachter ermittelt mittels im ruhenden Systeme aufgestellter, gemäß §1 synchroner, ruhender Uhren, in welchen Punkten des ruhenden Systems sich Anfang und Ende des auszumessenden Stabes zu einer bestimmten Zeit t befinden. Die Entfernung dieser beiden Punkte, gemessen mit dem schon benutzten, in diesem Falle ruhenden Maßstabe ist ebenfalls eine Länge, welche man als "Länge des Stabes" bezeichnen kann.

Let there be given a rigid rod at rest; its length, measured by a measuring rod that is also at rest, shall be *l*. We now imagine that the axis of the rod is placed along the X-axis of the coordinate system at rest, and that the rod is then set in uniform parallel translational motion (speed *v*) along the X-axis in the direction of increasing *x*. We now seek to determine the length of the *moving* rod, which we imagine to be obtained by the following two operations:

(a) The observer co-moves with the above-mentioned measuring rod and the rod to be measured, and measures the length of the rod directly, by applying the measuring rod exactly as if the rod to be measured, the observer, and the measuring rod were at rest.

(b) Using clocks at rest that are set up in the system at rest and are synchronous in the sense of §1, the observer determines the points of the system at rest at which the beginning and the end of the rod to be measured are found at some given time *t*. The distance between these two points, measured by the rod used before, which in the present case is at rest, is also a length, which can be designated as the "length of the rod."

There are two ways to measure the length of an object that is in motion with respect to the frame considered at rest. One is to run next to the moving object and compare it with the sample ruler. The other is to arrange a series of synchronized clocks along the X-axis, and read the time at which the front end of the moving object passes near a clock A, the time at which the back end passes near a clock B, and if the time marked by both clocks is the same, conclude that the length of the moving object is equal to the distance between clocks A and B. For example: a boat moves along a canal, and two observers on the shore having synchronized clocks take the time when they see one the bow and the other the stern. Then they tell each other the time they surveyed, and if they check that it's the same for both, they conclude that the boat is as long as the distance separating them.

Of course, such method is impractical: innumerable observers should be fitted with a clock in order to find two among them who attest to have seen one the bow and the other the stern when the clock marked the same time. But the text exposes the example in this way to simplify the situation to the fullest extent, and then infer consequences. We can soon imagine more advanced methods: observers could determine the speed of the boat with observations through their chronometers, and then take any time when the passage of the bow and the stern is observed, and develop simple calculations to know the length of the boat. But the focus of the argument is to point out how the determination of time of events serves to measure remote distances. From these principles technology has developed to radar and to today's laser telemetry used daily, but here are discussed the theoretical consequences of space measurements carried out through chronometers, using the example of a rudimentary archetype of this kind of measure.

§2, Paragraph 3.

Nach dem Relativitätsprinzip muß die bei der Operation a) zu findende Länge, welche wir "die Länge des Stabes im bewegten System" nennen wollen, gleich der Länge l des ruhenden Stabes sein.

According to the principle of relativity, the length to be found in operation (a), which we shall call "the length of the rod in the moving system," must equal the length *l* of the rod at rest.

Already according to the Galileo's law of inertia, but also according to the principle of relativity, which extends the law of inertia, the operation that measures a length by putting the ruler close to it and counting the repetitions of this act must give the same result for systems either at rest or in motion. If we measure a piece of wood with a ruler while we are at rest on the Earth we must get the same result we would get by measuring the same piece of wood travelling on a ship or a train.

§2, Paragraph 4.

Die bei der Operation b) zu findende Länge, welche wir "die Länge des (bewegten) Stabes im ruhenden System" nennen wollen, werden wir unter Zugrundelegung unserer beiden Prinzipien bestimmen und finden, daß sie von l verschieden ist.

We shall determine the length to be found in operation (b), which we shall call "the length of the (moving) rod in the system at rest," on the basis of our two principles, and will find it to be different from *l*.

This is a surprise announcement: the length of a moving object measured by means of time determinations and synchronized clocks will not yield the same result as the measurement performed directly by an observer who simply puts the metric sample near the object. This is not obvious, but it is the specific assertion of the theory we are exposing.

§2, Paragraph 5.

Die allgemein gebrauchte Kinematik nimmt stillschweigend an, daß die durch die beiden erwähnten Operationen bestimmten Längen einander genau gleich seien, oder mit anderen Worten, daß ein bewegter starrer Körper in der Zeitepoche t in geometrischer Beziehung vollständig durch denselben *Körper, wenn er in bestimmter Lage* ruht, *ersetzbar sei.*

The commonly used kinematics tacitly assumes that the lengths determined by the two methods mentioned are exactly identical, or, in other words, that in the time epoch *t* a moving rigid body is totally replaceable, as far as geometry is concerned, by the *same* body when it is *at rest* in a particular position.

Einstein says: what has been said in the previous paragraph is not obvious, and contradicts the universally used kinematics. The latter assumes implicitly that the measurement of the boat in motion made by observers at rest by clocks will give exactly the same result of the measure made by juxtaposing the ruler to the length of the boat. Even the readers of these lines are expected to agree with classical kinematics, if they do not already know Einstein's theory.

§2, Paragraph 6.

Wir denken uns ferner an den beiden Stabenden (A und B) Uhren angebracht, welche mit den Uhren des ruhenden Systems synchron sind, d. h. deren Angaben jeweilen der "Zeit des ruhenden Systems" an den Orten, an welchen sie sich gerade befinden, entsprechen; diese Uhren sind also "synchron im ruhenden System".

Further, we imagine that the two ends (A and B) of the rod are equipped with clocks that are synchronous with the clocks of the system at rest, i.e., whose readings always correspond to the "time of the system at rest" at the locations they happen to occupy; hence, these clocks are "synchronous in the system at rest."

In this paragraph, the reader is instructed to imagine two clocks A and B which are in the moving system and mark the same time, and that there are two observers near these clocks who watch the system at rest, and see all the clocks of the system at rest, which are many, passing under their eyes. Before starting, being at rest, the clocks A and B were synchronized with those of the system at rest using the procedure we know; after leaving, the two moving observers watch each their clocks A and B moving along with them, and watch together with their clocks the ones in the system at rest, which they meet while they move on. The two moving clocks and the ones at rest that are encountered during the journey all mark the same time, because all the clocks have been synchronized before, at rest, and it is assumed that all the clocks considered in the example work properly.

§2, Paragraph 7.

Wir denken uns ferner, daß sich bei jeder Uhr ein mit ihr bewegter Beobachter befinde, und daß diese Beobachter auf die beiden Uhren das im §1 aufgestellte Kriterium für den synchronen Gang zweier Uhren anwenden. Zur Zeit t_A gehe ein Lichtstrahl von A aus, werde zur Zeit t_B in B reflektiert und gelange zur Zeit t'_A nach A zurück. Unter

Berücksichtigung des Prinzipes von der Konstanz der Lichtgeschwindigkeit finden wir:

$$t_B - t_A = \frac{r_{AB}}{c - v}$$

und

$$t'_A - t_B = \frac{r_{AB}}{c + v}$$

wobei r_{AB} die Länge des bewegten Stabes — im ruhenden System gemessen — bedeutet. Mit dem bewegten Stabe bewegte Beobachter würden also die beiden Uhren nicht synchron gehend finden, während im ruhenden System befindliche Beobachter die Uhren als synchron laufend erklären würden.

Anmerkung: "Zeit" bedeutet hier "Zeit des ruhenden Systems" und zugleich "Zeigerstellung der bewegten Uhr, welche sich an dem Orte, von dem die Rede ist, befindet".

We further imagine that each clock has an observer co-moving with it, and that these observers apply to the two clocks the criterion for synchronism formulated in §1. Suppose a beam of light starts out from A at time t_A, is reflected from B at time t_B, and arrives back at A at time t'_A. Taking into account the principle of the constancy of the speed of light, we find that

$$t_B - t_A = \frac{r_{AB}}{c - v}$$

and

$$t'_A - t_B = \frac{r_{AB}}{c + v}$$

where rAB denotes the length of the moving rod, measured in the system at rest. The observers co-moving with the moving rod would thus find that the two clocks do not run synchronously while the observers in the system at rest would declare them synchronous.

Footnote: "Time" here means both "time of the system at rest" and "the position of the hands of the moving clock located at the place in question."

Here the expression has become too concise to be clear, and we must explicitly develop calculations so that the text assumes a definite and unambiguous meaning.

The previous paragraph placed on the moving system two clocks A and B and two characters near these clocks. Let us now call M the moving system, and F the system at rest, to avoid repetition of the expression. The characters in motion decide to repeat the synchronization procedure by sending a beam from A to B and reflecting it from B to A. The beam is emitted from point A on the moving rod, arrives at point B on the moving rod and finally returns to point A on the moving rod.

The action is observed from both the point of view of whoever is on the moving rod, and from the point of view of whoever is in the system at rest. In the frame of the moving rod, the beam goes from A to B and returns from B to A running twice the same distance. That is, from the point of view of the moving system:

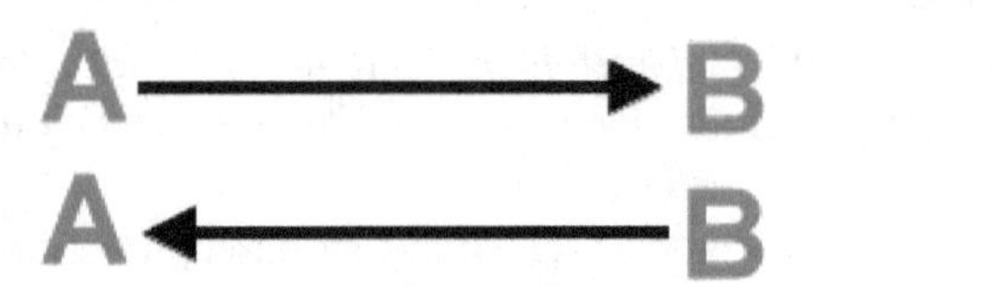

The beam goes from A to B and returns from B to A through equal segments. The synchronization procedure is valid.

From the point of view of the system at rest instead the beam goes from A to B while point B escapes to the right, and then returns from B to A while point A meets it. So these three states follow one after another:

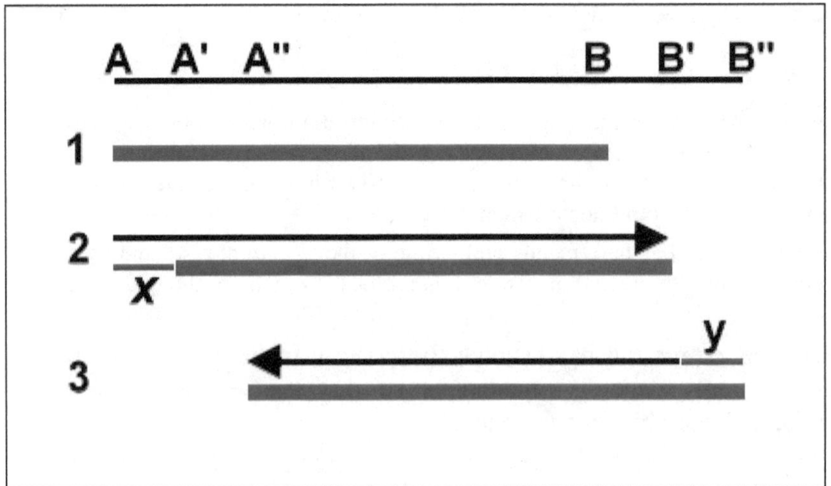

- State 1, Initial: the moving system is between A and B. The light beam starts from A in the direction of B.

- State 2: the beam reached clock B, which in the meantime moved to B' because all the moving system moved to A'B' at speed v. The beam travelled distance AB', which is equal to a length x plus the length of the moving rod.

- State 3: the beam started from B' in the direction of A' but while the moving system moved to A"B" at speed v, so clock A came towards the beam and the beam reached clock A in A". This time the beam travelled a distance equal to the length of the moving rod less a length y.

We now need to consider the time needed for the journey forth and back. It is here that the principle of absolute constancy of the speed of light gives rise to particular evaluations, so first we shall see what would happen without adopting the principle of absolute constancy, and then we shall see what happens by adopting it.

In both cases we set:

x=length of the segment to be added to the length of the moving rod in the forward beam travel

y=length of the segment to be subtracted from the length of the moving rod on the return beam travel

and keeping the notation of Einstein's text we set:

r_{AB}=moving rod length

t_B-t_A=time needed for the forward journey

t'_A-t_B=time needed for the return journey.

Let us first consider the speed of light relative to the source — do not let us assume the postulate of absolute constancy and consider that the speed of light is to be added to that of the source — and let us study the situation under this hypothesis.

On the way forth, we have:

$$x = (tB - tA)v$$

because clock A has travelled the distance x from A to A' at speed v, and we have:

$$rAB + x = (tB - tA)(c + v)$$

because in the fixed system, if we do not assume the constancy of the speed of light, the signal moves at speed $c+v$.

Hence replacing x we find the forward journey time:

$$rAB + (tB - tA)v = (tB - tA)c + (tB - tA)v;$$
$$rAB = (tB - tA)c;$$
$$tB - tA = \frac{rAB}{c};$$

On the way back, we have:

$$y = (t'A - tB)v$$

because clock B has travelled the distance y from B' to B" at speed v, and we have:

$$rAB - y = (t'A - tB)(c - v)$$

because in the fixed system, if we do not assume the constancy of the speed of light, the return signal moves at c-v.

Hence replacing y we find the return time:

$$rAB - (t'A - tB)v = (t'A - tB)c - (t'A - tB)v;$$
$$rAB = (t'A - tB)c;$$
$$t'A - tB = \frac{rAB}{c};$$

Thus, assuming that the speed of light must be added to the speed of the source, t_B-t_A and t'_A-t_B have both the same value r_{AB}/c, and therefore the synchronization procedure carried out by the moving system is valid also from the point of view of the system at rest. In fact, the forward journey time is the same as the return time.

Let us now consider the speed of light a constant — that is, assume the postulate of absolute constancy — and let us look at the situation under this hypothesis, which is the final one of Einstein.

On the way forth, we have:

$$x = (tB - tA)v$$

because clock A has travelled the distance x from A to A' at speed v, and we have:

$$rAB + x = (tB - tA)c$$

because in the system at rest, assuming the constancy of the speed of light, the signal moves at speed c as in the moving system and in each other.

Hence replacing x we find the forward journey time:

$$rAB + (tB - tA)v = (tB - tA)c;$$
$$rAB = (tB - tA)c - (tB - tA)v;$$
$$rAB = (tB - tA)(c - v);$$
$$tB - tA = \frac{rAB}{c - v};$$

On the way back, we have:

$$y = (t'A - tB)v$$

because clock B has travelled the distance y from B' to B" at speed v, and we have:

$$rAB - y = (t'A - tB)c$$

for the same reason as the way forward.

Hence replacing y we find the return time:

$$rAB - (t'A - tB)v = (t'A - tB)c;$$
$$rAB = (t'A - tB)c + (t'A - tB)v;$$
$$rAB = (t'A - tB)(c + v);$$
$$t'A - tB = \frac{rAB}{c + v};$$

Therefore, in the second case, that is, under the assumption of the absolute constancy of c, $t_B\text{-}t_A$ and $t'_A\text{-}t_B$ DO NOT have the same value, and therefore the synchronization procedure performed by the moving system is NOT valid also from the point of view of the fixed system. Since the situation is reciprocal, synchronization is considered invalid. The text literally says that it is the system in motion which detects that synchronization is not valid, but the development of reasoning that justifies the algebraic expressions of the text is precisely valid from the point of view of the system at rest. However, this does not matter because on account of the law of inertia (and the principle of relativity) both systems give the same judgment about each other.

At this point, there is an important remark to be made. As we can see from the argument, Einstein assumes that light moves not with respect to the ether but with respect to the emitting body, as in the emission theory. This is not explicitly stated, but it is implied, because without this assumption what is argued in paragraph 7, and what we have long analysed, would make no sense. In paragraph 7, the light starts at the extreme A of the moving system at speed c, reaches the point B of the moving system and is reflected to return to A. The movement is relative to the moving system. The observer at rest sees the moving system and sees the light in motion within the moving system: the light source in A, the beam and the reflecting mirror B for the observer move together with one another. The peculiarity of Einstein's hypothesis is that the speed of light is c, even if measured by the system at rest, and this gives rise to the difference in temporal determinations between the two systems, which would not exist if the emission theory were assumed, and hence if the system at rest attributed the speeds $c+v$ and $c-v$ to the beam moving through A and B.

The assumption that light propagates with respect to the source, though not at a speed equal to its own added to that of the source, and therefore the partial assumption of the emission theory, is implicit in all the exposures of the central principle of special relativity, which are all developments and extensions of what is stated in this paragraph of ED1905. In particular, this assumption is evident in the use of the so-called "light-clock", a typical auxiliary tool to explain the theory. The light-clock consists of a theoretical apparatus in which a beam is emitted by a source at point A and is reflected by a mirror in B, so that the beam first travels AB and then BA. The figure is simply this:

The complete AB+BA journey is assumed as a unit of time, and therefore the apparatus is a clock. While for the observer at rest the beams emitted by the light-clock travel the distance AB as they go forward and return, for an observer in motion with respect to the clock the beam would move this way, describing a sawtooth trajectory:

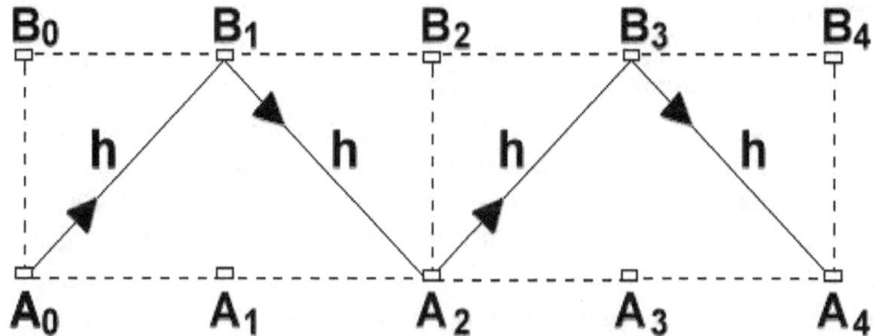

The beam started from source A in A_0 would reach mirror B when the mirror moved to B_1. Then the reflected beam would reach the source A when A moved to A_2, and so on. In this way, the beam would not travel distance AB, but the distance $h=A_0B_1$, which is the hypotenuse of the triangle that has the catheti A_0B_0 (that is AB) and B_0B_1 (equal to A_0A_1). Distance B_0B_1 depends on the speed at which the observer moves with respect to the clock, and the hypotenuse h is greater than the distance AB. In the clock frame the beam obviously moves at speed c. If the speed of the light were relative to the source, for the moving observer the beam would move at the speed x proportional to the hypotenuse, and we would have $c{:}AB=x{:}h$, hence $x=(h{\times}c)/AB$. But in the hypothesis of relativity, the speed of light cannot be greater than c for anyone. Thus, from the example of the light clock follows that the space and time measures of the AB system change for the observer in motion so as to preserve the speed of light c for all the observers. This kind of argument is a variant of that developed in ED1905, and adds nothing to it. We have mentioned the light-clock just to point out that in special relativity it is always assumed that light moves together with the emitting body, even though it does not inherit its inertia and speed. If there were no such tacit assumption, neither the arguments of ED1905 nor the subsequent variations similar to that using the light-clock could be developed.

Let us go back to the paragraphs in the second chapter of ED1905.

§2, Paragraph 8.

Wir sehen also, daß wir dem Begriffe der Gleichzeitigkeit keine absolute
*Bedeutung beimessen dürfen, sondern daß zwei Ereignisse, welche, von
einem Koordinatensystem aus betrachtet, gleichzeitig sind, von einem
relativ zu diesem System bewegten System aus betrachtet, nicht mehr als
gleichzeitige Ereignisse aufzufassen sind.*

Thus we see that we must not ascribe absolute meaning to the concept of
simultaneity; instead, two events that are simultaneous when observed
from some particular coordinate system can no longer be considered
simultaneous when observed from a system that is moving relative to
that system.

Given all what was said above, these conclusive words are
justified. It is left to the reader the further conclusion that length
measurements by means of clocks, such as those described in
paragraph 2, will no longer be absolute but will become related to the
state of motion, in the same way as judgments on simultaneous
events and duration measurements.

However, we have to make some observations regarding the
interpretation of the text, which leaves us now legitimate doubts. I
put myself in the shoes of readers who have just heard that clocks in
motion disagree with clocks at rest and readers who know these
things from relativistic literature, but admit that they did not
understand them. It is not to be thought that trying to answer these
problems using the original text is an antiquarian erudition exercise,
and that it would be better to study the problems using more recent
treatises, because the textbooks of every epoch over the more than
one hundred years that have passed neither correct the theory, nor do
they ever adopt views other than the 1905 version. Interpretative
problems exist, as evidenced by the fact that there is a considerable
number of books that say the same things, and there are many people
who have tried to study the theory and have decided to accept it
without having grasped it, because they are persuaded by the general
consensus the theory receives. The literature after Einstein proceeds
along the same lines of Einstein's exposition, and is unable to resolve
the original ambiguities in interpretation. So there is no reason to
refer to recent textbooks: additional interpretation problems would be
added to those inherent within the original text, which, if anything,
has the advantage of brevity.

3. Einstein's solution

In the sixth paragraph, Einstein insists that the clocks of the two systems in an initial state may be synchronous and indicate the same time. To make the text understandable, we imagine that the clocks were synchronized with the procedure described above when everything was at rest, but to point this out is definitely not necessary: there may be an initial state in which the clocks mark all the same time, and if the clocks do not go wrong they will continue to mark it.

However, due to the absolute constancy of the speed of light, there is a problem of synchronization getting out of phase, because there is a situation in which two times considered equal with right in the moving system are also considered different in the system at rest with the same right. What does this mean? What happens to the clocks?

There are two possible answers. The first answer is that a second synchronization procedure could be defined, which has no description in the text, which instead of being defined for clocks at rest, should be defined for clocks in relative motion. After executing this procedure, moving clocks that earlier marked the same time would mark different times due to the constancy of the speed of light. We shall call this procedure the hypothetical synchronization-in-motion procedure, because it is not explicitly described either by Einstein or by relativistic literature, but an implicit reference to it is made in the third chapter of ED1905, and we shall see it briefly. The second answer, on the other hand, might be that the two systems moving with respect to each other are related to each other by relationships that define a kind of contraction or dilatation of times and spaces of the one with respect to the other, and in this case the clocks should be able to run at different rates for natural, physical reasons, and not following a dedicated synchronization procedure implemented by the observers. Knowing Lorentz's hypothesis, we can conceive that the answer could be this, but the reader should realize, or have the presentiment, that the situation is not the same as described by Lorentz: there is no ether here that allows us to conceive a contraction of the systems that move in the sense of the ether wind, and instead there is the postulate of the principle of relativity that says that simple speed in uniform rectilinear motion does not have physical effects.

3. Einstein's solution

We shall soon see, coming to electrodynamics, that in the text of 1905 the answer is none of them, but perhaps everything is simpler. But given the interpretation that the theory has received over the subsequent years, the question cannot be ignored and finding the answer becomes indispensable. However, we shall also see that relativistic literature is unable to answer this question, and gets away with it repeating infinite times a certain formula, which sounds: "the clocks run at the same rate, but the time of the moving system is dilated." This formula of relativistic literature might be called the relativistic principle of Ptolemy: they are not the lengths that contract, it is space that dilates. They are not the movements and changes that slow down, it is time that dilates. It is not the Earth that revolves around the Sun, it is the Sun revolving around the Earth. The expression declined in many variants, "the clocks remain the same, but the time of the moving system is dilated" means nothing, does not answer the question, and it is just stupid: because if time dilates in a given frame, then all the phenomena inside it slow down, and the clocks present there must indicate different times than those installed in frames where time does not dilate. The two notions, slowing down of duration of physical processes, and time dilatation, are reciprocal.

3.9   *ED1905 Chapters 3, 4 and 5, Lorentz transformations and consequences*

Now it is no longer necessary to follow the text word by word. We can be more concise without compromising understanding. In Chapter 3, Einstein asks: if we have two systems in relative motion and if in them the speed of light is the same, what will be the relationship between space and time magnitudes and coordinates of the same events as seen in the two systems? Asked this question, Einstein uses an algebraic procedure he develops in a form that is usually not followed by subsequent authors, and comes to the results we know: the Lorentz gamma factor and Lorentz transformations. From a mathematical standpoint, the situation is the same (or more exactly, is analogous): we have systems in relative motion in which the speed of light is the same. What is the relationship between measurements in the two systems? The different physical interpretation of the situation is not relevant to this relationship, and so Einstein achieves identical results to those of Lorentz. Even the fact that he gets them through an algebraic procedure different from those used in the past is irrelevant.

Then, in Chapter 4, Einstein illustrates the consequence that if the movement takes place along the x-axis, after applying Lorentz's formulas an object which is spherical in a frame, would look like an ellipsoid in another: a simple consequence, discussed previously also by Lorentz and Poincaré.

Then, again in Chapter 4, we find an important statement[24]:

> *Wir denken uns ferner eine der Uhren, welche relativ zum ruhenden System ruhend die Zeit* t, *relativ zum bewegten System ruhend die Zeit* t' *anzugeben befähigt sind, im Koordinatenursprung von* K' *gelegen und so gerichtet, daß sie die Zeit* t' *angibt. Wie schnell geht diese Uhr, vom ruhenden System aus betrachtet?*

> We further imagine that one of the clocks that is able to indicate time t when at rest relative to the system at rest and time t' when at rest relative to the system in motion, is placed in the origin of K' and adjusted so that it indicates the time t'. What is the rate of this clock when observed from the system at rest?

---

[24] ED1905, § 4.

What speed has the moving clock with respect to the clock at rest? The answer is simple: it has the speed corresponding to what we get from Lorentz's formulas. Einstein expresses this by calling this an "eigentümlich" consequence, translatable as "peculiar," or "remarkable"[25]:

> *Hieraus ergibt sich folgende eigentümliche Konsequenz. Sind in den Punkten A und B von K ruhende, im ruhenden System betrachtet, synchron gehende Uhren vorhanden, und bewegt man die Uhr in A mit der Geschwindigkeit* v *auf der Verbindungslinie nach B, so gehen nach Ankunft dieser Uhr in B die beiden Uhren nicht mehr synchron, sondern die von A nach B bewegte Uhr geht gegenüber der von Anfang an in B befindlichen um 1/2 tv²/c² Sek. (bis auf Größen vierter und höherer Ordnung) nach, wenn* t *die Zeit ist, welche die Uhr von A nach B braucht.*

This yields the following peculiar consequence: If at the points A and B of K there are located clocks at rest which, observed in a system at rest, are synchronized, and if the clock in A is transported to B along the connecting line with speed v, then upon arrival of this clock at B the two clocks will no longer be synchronized; instead, the clock that has been transported from A to B will lag *1/2 tv²/c²* seconds (up to quantities of the fourth and higher orders) behind the clock that has been in B from the outset, if t is the time needed by the clock to travel from A to B.

For some reason, not easy to interpret, the expected result is expressed in approximate form, but this only depends on how the calculation is developed. It is important that the formulation of question and answer contains both a problem and a very strong interpretative indication, since it is explicitly stated that a clock can delay as a result of the relative motion condition, but this is said with respect to the case of a moving clock "so gerichtet, daß sie die Zeit t' angibt", that is "adjusted so that it indicates the time t'". Hence, the text speaks of an act in which someone has set the moving clock in such a way as to indicate the time t'. Thus, reference is made to the existence of a procedure, which is never explicitly described, by which clocks in motion would be synchronized with a voluntary act carried out by human observers, and as a result of which they would indicate different times. Everything said by Einstein earlier should therefore be interpreted in the sense that if this intentional procedure

---

[25] ED1905, § 4.

is not performed, then the clocks that have been synchronized at rest once started will preserve synchronism, because there is no natural reason why they should mark the time transformed according to Lorentz. So, are we talking about contractions of lengths and durations, or dilatations of space and time, which are real, and which depend on physical reasons, or which are apparent and depend on conventions? Here it would seem that we are talking about mere conventions, but the application to electrodynamics will immediately give us a different answer to this question.

It should be further pointed out that shortly thereafter Einstein makes a strange assertion, which has been the subject of discussions. Einstein says that a clock installed at the equator would delay with respect to a clock installed on one of the Earth poles, and specifies that the circular travel of the clock at the equator can be interpreted as the sum of infinite straight lines that make up a polygonal chain, and therefore the Lorentz formulas can be applied to the motion of the equatorial clock. In many textbooks this clarification is understood as explicitly asserting that the delay of the equatorial clock is due to physical causes and does not depend on synchronization procedures. But those who interpret it as so did not read carefully the previous lines of the same paragraph, and did not observe the expression "so gerichtet, daß sie die Zeit t' angibt", or "adjusted so that it indicates the time t'". This expression referring to clocks that delay, indicates that the delay stems from a synchronization procedure. The final example of Einstein is merely illustrative, and it means that if an equatorial clock were synchronized with the known procedure, then it would retard compared to that of the pole. The example has given rise to misunderstandings simply because it is badly chosen and written: to have a meaningful example we should imagine having something like a counter-rotating system with respect to the equator and synchronizing the equatorial clock with a clock in this system and imagine that the polar clock, on the other hand, is synchronized with the clock in the counter-rotating system: so the equatorial clock would delay compared to the polar one (I leave readers the exercise to develop this correction if they want to go into detail).

3.10   Review of previous sections

At the beginning of this chapter, I warned that the sections up to the present would be the most difficult to read throughout the book. Since this book was written for beginners and the perplexed, I guess that at this point readers of both categories are confused, and the perplexed ones are confirmed in perplexity. The reason is this: we have read the heart of the exposition of a theory which everyone knows to be considered unanimously a masterpiece of twentieth-century scientific thought, a victory of reason on the limits and mental restraint of classical physics. The intelligent readers, but inexperienced, therefore expected to find a solid and consistent theory, and instead they found some indications that gave them an idea of how one could get to a certain result, but who are still so vague not to allow them to see the work done and the result achieved. Let us say it is too early to pretend to thoroughly understand and grasp the theory. The puzzled readers now only need to make sure that they grasped the meaning of the pages they have read without misunderstanding them, and should not yet expect to know how to compose them in a complete and consistent theory.

Let us summarize what we have read so far. We have seen, and repeated too many times, that in Lorentz theory c is an absolute constant because the problem of the observer's speed with respect to the source of electromagnetic waves does not exist, and the observer's speed with respect to the ether is hidden by the phenomenon of Lorentz contraction. We have seen that there are many reasons for considering the light moving not in the ether, but in the vacuum with respect to the bodies that emit it, and that in this case the problem of speed with respect to the ether does not exist by definition, but obviously there would be the problem of speed with respect to the light source, which should always be added to c. In order to maintain c an absolute constant, Einstein has proposed to assume by hypothesis that c is constant for every observer at rest or in motion with respect to the source of the waves, and has promised to be able to find a model of nature within which this hypothesis, at first sight incompatible with the law of inertia and relativity principle (as observed by Einstein himself), could become conceivable.

3. Einstein's solution

At this point the readers are not (yet) able to get an idea of how this can be explained. The readers thinks of themselves moving away or approaching any source of light, and cannot help thinking that if there is no ether, and if light is conceived as a sort of sequence of bullets emitted by its sources, then the speed of light with respect to the observers will always be c plus or minus their speed with respect to the light source. It seems impossible to conceive a kinematics corresponding to Einstein's hypothesis, because if there is no universal and common frame of the kind of ether, then if I move for example at speed 10 with respect to a light source and at speed 100 with respect to another, I cannot be subject to a contraction of the kind of Lorentz's that nullifies both the speed 10 of the first light source and the speed 100 of the second, and also all the countless other speeds at which I move with respect to the countless light sources that may exist. The assumption of a unique and universal reference frame (whether it be the ether or something else) is indispensable to conceive Lorentz's contraction. However, when verbally expressed, Einstein's postulate has a comprehensible meaning because it evokes two clearly defined situations. The postulate has two distinct semantic components: the first component claims that light moves like a bullet to its source, while the second assumes that those who timed a beam of light would always find the speed c, being either at rest or in motion with respect to the source. But, at least at first glance, it is impossible to put the two components together in a single rational and not fanciful representation, and therefore the postulate (at least for the moment) has the semantic status of a fairy tale. Let us imagine a small fragment of science fiction: from Coma Berenices, a spacecraft pilot flies at a hundred thousand metres per second and overtakes two slower ships moving in the same direction. These two ships move one at ten thousand and one at twenty thousand metres per second with respect to the same origin of the motion than the first ship, but, for some reason inherent in the intrigue of the tale, the author decides that the fastest ship pilot moves at the speed of eighty thousand metres per second with respect to both the ships that he overcomes, instead of ninety and eighty thousand respectively. Is this possible in nature? No. But does the discourse have semantics? Yes, because fairy tales have a fantastic sense consisting in the images we are free to create in our mind: even the story of Cinderella's coach that will turn into a pumpkin at midnight is not without semantics, but has a fantastic sense perfectly clear and defined. Einstein's second postulate has not passed this condition so far.

But as far as the case of light is concerned, Einstein has proposed that he is able to solve the problem of the second postulate with an argument not equal, but structurally similar to that of Lorentz, and that in fact leads to the same formulas for the contraction of durations and lengths. What did Einstein say to build this explanation? And what does relativistic literature say, expanding the original exposure without ever adding any new element to it?

Einstein described the theoretical model of a rudimentary machine to synchronize two distant clocks with electromagnetic signals. This rudimentary machine would not be properly designed if there were the ether, while it would be properly designed in the case of ballistic theory, where the speed of electromagnetic waves is relative to the source. Then he pointed out that under the assumption that c is constant even in the ballistic model, this machine would also give rise to the same measurement differences as would have occurred in the model of the ether, and which can be calculated with the Lorentz's formulas.

Then Einstein exposed the theoretical model of a rudimentary machine to measure lengths of moving objects by sending electromagnetic signals and by time measurements. This rudimentary machine is the archetype of today's laser telemeters. Concerning this machine, the same considerations as were made for synchronization are valid: this rudimentary machine would not be properly designed if there were an ether, it would be designed correctly in the case of ballistic theory, but under the assumption that c is an absolute constant even in the ballistic model, it would give rise to the same measurement differences as would occur in the ether model, which can be calculated by Lorentz.

Then Einstein hinted that as a result of some unspecified synchronization procedure, which we can imagine, two identical clocks in relative motion may be able to mark time values correlated by Lorentz's formulas instead of equal values. Since the first hypothesis is that of the principle of relativity, according to which the condition of uniform rectilinear motion does not give rise to physical consequences and cannot be distinguished from the state of rest, we think we have to interpret that two clocks that work well and mark the same time will continue to mark it, if one of them moves on a straight line at constant speed. Why should it happen otherwise, under the hypothesis of the law of inertia extended into the principle of relativity? In fact, when Einstein alludes to a clock that marks the time transformed according to Lorentz, he speaks of a clock "adjusted" in this way by a voluntary procedure, not of a natural phenomenon.

All this said, and recapitulated in this way that we have read before, can readers understand how it is possible that c is an absolute constant even in the absence of the ether? The answer is categorically no. We have not been told how Einstein's second postulate is possible. We have been told that if we take Einstein's second postulate, and if someone takes measures by applying the rules that would be valid in the simple ballistic theory, then this observer would have erroneous measures, and differences could be calculated with Lorentz's formulas. With this, we have no element to understand either how the second postulate can be possible, or what the usage should be of the theory that is outlined: some people might make mistakes in measurement because they do not know the second postulate, which, however, continues to look like the gimmick of a science fiction story, not a science principle. So? What changes in the world for this? Nothing, and in particular, running clocks continue to mark the same time as those at rest, because the principle of relativity prescribes that.

Putting it in a schema, what we have been told by Einstein is reduced to very little:

1. Let us assume that there is no ether and yet c is constant for every observer at rest or in motion with respect to the luminous source;

2. Now we shall explain how this is possible;

3. We remark that a moving observer who makes remote measurements by applying the rules valid in the case that the propagation took place according to the ballistic hypothesis, would obtain values differing from those corresponding to the hypothesis of point 1;

4. The measurement differences would be those of Lorentz's formulas;
5. So it follows ....

Nothing logically follows. Nothing has been said that makes statement 1 comprehensible to us, and nothing to fulfil the purpose stated in statement 2. Statements 3 and 4 do not show nor explain anything, being only descriptions of one of the endless situations that may arise. Other observers could carry out remote measurements by applying different criteria following countless hypotheses, and obtain measurement differences by coefficients of any kind instead of the results according to Lorentz.

So at this point the readers must accept the fact that they can attribute an exact and unambiguous meaning to the words spoken so far, but that they have not grasped the rational connection of the elements of the theory that Einstein is building. For those who have read carefully what has been said so far, the result cannot be different from this. Those who thought they understood the theory would certainly possess other elements, which we have not mentioned yet. Instead, those who understand the meaning of the different things they have read, but not their overall connection, should not be worried at all: these readers have understood exactly what could be understood.

3.11 ED1905 Chapter 6, application to electrodynamics and invariance of Maxwell equations

Some elementary notions are sufficient to understand the relevance of the theory of special relativity in relation to the issue of electrodynamics, because reasoning does not apply only to electrodynamics, but to any case, even simpler, where we could have the problem of the presence of constant c in a physical law. We do not need to get into the details of electromagnetic laws, but just know what they are.

We know that electric currents and magnetic phenomena exert mutual forces. If we let an electric current pass near a compass the needle deflects, if we move a magnet near a circuit built in a certain way then in this circuit an electric current is generated, and if we let current pass within a circuit built in some other way then this assumes magnetic properties. These elementary phenomena correspond to technologies that everyone knows: the force exerted by currents and magnets is the force that drives electric motors, the ability of a magnet to generate a current is exploited by dynamos and current generators, the possibility of transforming an electrical circuit into a magnet is the principle of operation of electromagnets, which have magnetic properties or have not according to the state of a circuit, and are therefore used to move iron scrap, as we see, for example, where old cars are lifted by an electromagnet to be then transformed into cubes by a press.

Electric circuits and magnets exert forces in the space surrounding them according to well-known laws. The spaces defined by these forces are called electrical and magnetic fields. The interaction between them is based on the mutual position of the electric and magnetic fields, depending on the distances and angles between them. There are two physical quantities, called "divergence" and "curl", which describe the behaviour of the electric and magnetic fields and that correlate the x, y, z coordinates of points of the electric and magnetic fields with the course of phenomena over time, i.e. with the coordinate t.

There are four fundamental equations, Maxwell's equations, which here are given just for curiosity in a very coarse form, not really algebraic, but just to give an idea. If E is the electric field and B is the magnetic field, we have:

$$curl\ (E)=\Delta B/\Delta t$$

$$divergence(E)=qv$$

$$curl(B)=\Delta E/\Delta t+i$$

$$divergence(B)=0$$

What does it mean? Something very accurate and complex. But we only need to know that divergence and curl are functions in which the constant speed of electromagnetic waves and the spatial and temporal coordinates of the electric and magnetic fields are introduced, so they are functions $f(c,x,y,z,t)$. Let us focus on:

$$divergence\ (B)=0$$

which is the simplest of the four equations, and says that the divergence (whatever it be) of a magnetic field is always zero. Now, this happens: the function $f(c,x,y,z,t)$ returns zero for certain values of x, y, z and t. So it tells us the groups of values x, y, z and t that we can find in a magnetic field. This represents some real physical properties of electromagnetic systems, electrical equipment, radios, and so on. Now imagine placing equipment on a cart and moving it right along the x-axis. If there are no other electric and magnetic fields then those we consider, for the law of inertia we expect the equipment to remain the same and to work as before, because the state of rest and motion are indistinguishable and relative, and therefore they do not have to cause any change in the equipment state. If I put a radio or a dynamo on an electric traction train, I can expect that their behaviour changes in relation to the speed of the train, since high power currents are used to make the train move and therefore there are specific environmental electromagnetic conditions. But if I put a radio or a dynamo on a cart pulled by a horse or a steam train, I do not expect anything to happen. The state of the radio or the dynamo either at rest or in relative motion should remain the same. To better clarify the problem, let us formulate the example by observing that a radio and a dynamo can be considered at this moment at rest in relation to the Earth, but in motion at different speeds compared to Mars and Jupiter. The choice to assign to the electric apparatus speed zero with respect to the Earth or one of the speeds differing from zero with respect to the planets is arbitrary and depends on the point of view, that is, by the fundamental coordinate frame we choose for reasons of simplicity of our calculations; so it is obvious that the changes in our choice do not produce physical changes of the object we observe, and it is obvious that thinking the opposite would be almost a manifestation of magical thought, once the absolute frame of the ether has been eliminated.

But the form of electromagnetic equations is inconsistent with this principle. Considering *divergence(B)=0*, the *f(c,x,y,z,t)* function of the magnetic field divergence has mathematical properties such that if we consider the system moving at speed *v* with respect to one at rest and apply the Galilean transformations by replacing *x* with *x-vt* and then calculating *f(c,x-vt,y,z,t)*, then the result of the calculation is no more zero. But the calculation must give zero, since the system does not change, being only considered by us in motion. To accept that the calculation does not give zero is equivalent to denying the law of inertia for an electromagnetic apparatus, and to assert that the behaviour of an electromagnetic apparatus changes as a result of speed. But speed is always related to an arbitrarily chosen frame. So, how to accept the formula of the divergence for the magnetic field, which gives different results for different speeds, when it should always give zero, regardless of the speed, i.e. regardless of the possible translation of the magnetic field along an axis? The problem, which obviously covers all four Maxwell equations, is expressed by saying that Maxwell's equations *are not invariant under a Galilean transformation.*

The nineteenth-century physicists looked at the problem in another perspective: they had conceived the theory under the undisputed assumption of ether, and therefore from their point of view it would have been necessary to determine the speed of the Earth and everything else with respect to the ether to achieve total accuracy. The application of the formulas without considering the speed with respect to the ether, from their point of view gave rise to an approximate result, but with a tiny and unobservable error given the magnitude of the speed of light with respect to the other speeds. Most important, this did not imply any inconsistency of the theory. By abolishing the ether, and considering the magnetic waves simply capable of propagating in the vacuum or in any environment, approximation in calculations remains the same, irrelevant, but the theory appears to be misleading from a conceptual point of view. The reason for everything is that in formulas there is the constant c, and the problem is what we discussed earlier, that no speed should ever be considered a physical constant if there is no unique universal frame to define it.

Now, the $f(c,x,y,z,t)$ function of the magnetic field divergence has mathematical properties such that if we replace both x and t transformed according to Lorentz instead of replacing only x by Galileo, then the result of the function calculation returns again zero, as it should be.

That is, if we have four values of x,y,z,t for which:

$$f(c, x, y, z, t) = 0$$

then the Galilean transformation does not give zero:

$$f(c, x - vt, y, \dot{z}, t) \neq 0$$

and instead the Lorentz transformation gives zero again:

$$f\left(c, \frac{x - vt}{\sqrt{1 - \frac{v^2}{c^2}}}, y, z, \frac{t - \frac{vx}{c^2}}{\sqrt{1 - \frac{v^2}{c^2}}}\right) = 0.$$

This happens for all four equations, and it is expressed by saying that Maxwell's equations *are invariant under a Lorentz transformation*. The mathematical demonstration is very complex; those who are curious may find it by consulting higher level textbooks. But be careful, partial or complete demonstration in any case only shows this: that a particular calculation that at rest gives zero, if Galilean transformations are applied gives no longer zero, while applying Lorentz's transformations it gives again zero. The mathematical demonstration by definition cannot prove anything other than this, and it cannot show, in particular, that there are physical realities that behave according to Lorentz's transformations.

Let us consider the problem from the nineteenth century point of view, that of Faraday, Maxwell and Lorentz. These physicists built electromagnetic theory by interpreting certain experiments on the assumption that there was ether, and that electromagnetic waves were a movement of the ether, as the waves of the sea are a movement of water. The hypothesis was not generic, but extremely detailed: the hypothetical characteristics of the ether were determined in a precise manner, according to several competing models, but no experimental checks could be made on the nature and existence of the ether. However, some equations were found which strictly accord with the electrical phenomena we experience. These equations are called Maxwell's for his fundamental contribution to their formulation. Now, how did the physicists of the nineteenth century interpret the problem of invariance?

The equation *divergence(B)=0* for them was valid in the ether. So for them there was a function of the divergence of the magnetic field *f(c,x,y,z,t)* that gives zero only for certain values of *t* and for certain values of *x,y,z* in the ether frame. In the nineteenth century they said: if we have an electromagnetic phenomenon that takes place on the Earth, and we have a time *t*, and we want to know the coordinates of a point of magnetic field point, then acquiring data measured on Earth and calculating we can find groups of values *x,y,z* with which the function gives zero, but since the Earth is in motion with respect to ether we made a mistake, because we should find values *x', y', z'* that correspond to coordinates of magnetic field points in the ether rather than on the Earth. Then, by admitting the ether, the physicists at that time did not have an interest in the problem of the equation invariance with respect to the translation of the coordinate system, but the problem was to determine the Earth's motion with respect to the ether, knowing its direction, sense, and speed. And having not found a way of knowing the motion of the Earth, they concluded (which was correct given their premise) that the application of Maxwell's laws always contained a mistake, albeit very small and irrelevant to practical applications: this mistake would have the magnitude of the ratio of the Earth's speed in ether to the speed of light. So the magnitude of the mistake would be insignificant, as insignificant are the numerical values of the corrections that are introduced by applying Lorentz's transformations. But the abandonment of the hypothesis of ether, recommended by the experiments made to find it and all failed, gives rise to the problem of invariance: it is no longer simply a matter of accepting an error that cannot be corrected, but the theory appears now spoiled by a conceptual error.

This said, let us return to Einstein's text. Chapter 6 opens up by presenting a system of six equations expressing in the extended algebraic form the relationships between the electric and magnetic field described by Maxwell equations. Then Einstein applies Lorentz transformations to these six equations and shows the achieved invariance. It is remarkable that the application of Lorentz transformations is introduced with the words: "Wenden wir auf diese Gleichungen die in §3 entwickelte Transformation an...", i.e. "Let us now apply to these equations the transformations developed in §3..." and no comment word is added. Lorentz transformations are applied, and therefore electromagnetic theory records a progress in the direction of consistency, because its equations become invariant. And that is all about the text: we do not need further comments about its meaning.

But we cannot help but wonder: why, according to Einstein, the application of Lorentz's transformations is legitimate? Why can we apply Lorentz's transformations to the physical reality of the electric and the magnetic field? Chapters 1 and 2 of ED1905 have constructed a representation of things so that relative motion systems may look mutually transformed according to Lorentz's rules, but everything has been described as if the phenomenon had a conventional or apparent character. It seems therefore that there is still a need to justify the application of Lorentz's transformations with rational arguments. Is there such a necessity? This will be discussed briefly, but the answer is: yes, because if the time transformed according to Lorentz were only what appears to a remote observer and not that marked by the clock of the system being studied, if the phenomenon had a merely apparent character, then the replacement of parameter t with the corresponding parameter t' transformed according to Lorentz would replace the times marked by the clocks with times not marked by the clocks and calculated with a criterion devoid of physical meaning, and the same would happen for the lengths on the x-axis, which would no longer correspond to their stable and unambiguous measures. With this, the operation would have no value, and the consistency of Maxwell's theory would have been reached with an arbitrary and fantastic criterion, that would add a new error to the old one by increasing the approximation and without formulating any serious theory of waves behaviour in the absence of the ether. The operation would be as insane as the one made by a person who, in order to believe him or herself rich, invented a bank with him or herself the owner and attributed to him or herself a credit balance of an imaginary account.

3.12 *ED1905 Chapter 6, a detail about electrodynamics*

In this section we hint at a detail that is most important to understand how the fortune of relativity theory began, but is not necessary for the meaning of the theory. The reader can thus be satisfied with a rough understanding without compromising the comprehension of the whole. The full understanding of the technical reach of Einstein's quote is reserved for specialists, but all that we have said so far should allow each reader a rough understanding of this detail, clear enough to fully grasp the relationship with special relativity.

At the end of Chapter 6 of ED1905, a qualitative aspect of the electromagnetic issue mentioned in the introductory page is commented upon. The essay opens with words already cited, but now let us read the whole first paragraph of the Introduction. I no longer quote German, because here we are no longer interested in the weight of the individual words[26]:

It is well known that Maxwell's electrodynamics — as usually understood at present — when applied to moving bodies, leads to asymmetries that do not seem to match with the phenomena. Let us recall, for example, the electro- dynamic interaction between a magnet and a conductor. The observable phenomenon depends here only on the relative motion of conductor and magnet, while according to the customary conception the two cases, in which, respectively, either the one or the other of the two bodies is the one in motion, are to be strictly differentiated from each other. For if the magnet is in motion and the conductor is at rest, there arises in the surroundings of the magnet an electric field endowed with a certain energy value that produces a current in the places where parts of the conductor are located. But if the magnet is at rest and the conductor is in motion, no electric field arises in the surroundings of the magnet, while in the conductor an electromotive force will arise, to which in itself there does not correspond any energy, but which, provided that the relative motion in the two cases considered is the same, gives rise to electrical currents that have the same magnitude and the same course as those produced by the electric forces in the first-mentioned case.

---

[26] ED1905, Foreword.

3. Einstein's solution

It is here that Einstein masters what he says, and expresses it with certainty and clarity, because he knows the laws of electricity, which he had experienced as a boy, collaborating in the work of his father and uncle. The considered phenomenon is very simple, and the need to express it so as to avoid unnecessary distinctions is understandable by anyone: when a magnetic field and an electric conductor rotate one with respect to the other, there is an electrical current that is the same either in the case in which we consider the electric field at rest, or we consider the magnetic field at rest. So why does classical theory describe the same phenomenon in different terms according to the two cases? Not only are Maxwell's equations not invariant to Galilean transformations, but the entire classical theory, based on the assumption of the ether, defines as absolute and distinct certain characteristics of phenomena which instead experience shows to be relative and indistinguishable. There is no point in saying that "the magnet is in motion and the conductor is at rest" or the opposite in an absolute sense, when the effect of the relative motion of the magnetic with respect to the electric field is the same.

In the introductory section the problem was posed; now in Chapter 6 the fruit of the method used is collected. I now quote the whole conclusion of Chapter 6 because, even if its technical sense is not exactly understood by every reader, it helps anyway to understand how applying Lorentz transformations, if it were legitimate, would yield results that would be a success in electromagnetic theory[27]:

> By way of interpreting these equations, we shall add the following: Imagine a pointlike quantity of electricity whose magnitude, measured in the system at rest K, is "one," i.e., which, when at rest in the system at rest, exerts a force of 1 dyne on an equal quantity of electricity at a distance of 1 cm. According to the principle of relativity this electric mass is also of magnitude "one" if measured in a moving system. If this quantity of electricity is at rest relative to the system at rest, the vector (X,Y,Z) equals the force exerted on it by definition. If this quantity of electricity is at rest relative to the moving system (at least at the instant considered), the force exerted on it, and measured in the moving system, will equal the vector (X',Y',Z'). Hence, the first three of the above equations can be expressed in words in the following two ways:

---

[27] ED1905, § 6.

1. If a pointlike unit electric pole is in motion in an electromagnetic field, there will act on it, in addition to the electric force, an "electromotive force" which, if we neglect terms multiplied by the second and higher powers of v/c, equals the vector product of the speed of motion of the unit pole and the magnetic force, divided by the speed of light. (Old mode of expression.)

2. If a pointlike unit electric pole is in motion in an electromagnetic field, the force acting on it equals the electric force present at the location of the unit pole, which is obtained by transforming the field to a coordinate system that is at rest relative to the unit electric pole. (New mode of expression.)

Analogous propositions apply for "magnetomotive forces." We can see that in the theory developed, the electromotive force merely plays the role of an auxiliary concept, whose introduction is due to the circumstance that the electric and magnetic forces do not have an existence independent of the state of motion of the coordinate system.

It is further clear that the asymmetry mentioned in the Introduction when considering the currents produced by the relative motion of a magnet and a conductor, disappears. Questions as to the "seat" of the electrodynamic electromotive forces (unipolar machines) also become pointless.

So, here we have read that there are definitions that if expressed according to the "old mode" have a convoluted form, whereas expressed according to the "new mode" they have a simpler and more linear form. Not only, therefore, electromagnetic theory becomes rid of the problems of Maxwell's equations invariance, but it reaches unambiguous definitions without asymmetries which are both incompatible with the general assumption of the law of inertia and lacking significance in relation to experience.

3.13 *Assumptions and consequences of applying Lorentz's transformations to electrodynamics*

Chapters from 7 to 10 of ED1905 apply Lorentz's transformations to other specific problems, which we shall not analyse, because what we have seen is enough to clarify the character and development of the overall operation made by Einstein. In subsequent chapters, Einstein acts by considering as an acquired result that the application of Lorentz's transformations is legitimate in any field, as he has considered in relation to Maxwell's laws, and consequently proceeds in further discussions that are no longer relevant to the problem which we must solve, i.e. that of the legitimacy of the postulate of the constancy of the speed of light and the meaning of the foundations of the theory. If we understand why application to the case of Maxwell's laws is allowed, that will be valid for any other problem to which Lorentz's transformations can be applied.

We now have elements to understand the stages of Einstein's thinking process, which we can describe schematically in these steps:

- There is a fundamental problem: abandoning the hypothesis of the ether, we must solve the problem of Maxwell's equations invariance. The problem has little or no practical meaning, but electromagnetic theory is not well formulated if it is not solved.

- To correct the form of Maxwell's equations, we should consider the whole problem, and no one has ever been able to.

- But if we apply Lorentz's transformations, the equations become invariant without changing form. The result is achieved.

- However, applying Lorentz's transformations we make an arbitrary act: we replace the times marked by the clocks with unjustified time measurements, which do not correspond to the direct measure, and replace the spaces measured with the ruler with unjustified space measures, also not corresponding to the direct measure.

- However, if space time measurements were relative to inertial systems and dependent on relative speed, we would have a perfectly accurate result.

- Therefore, we assume that light speed is constant in all inertial systems, that light travels at the same speed in inertial systems moving with respect to one another, that is, we postulate, for example, that a sunlight beam has the same speed for me, for a car that moves away from me, and for a car coming towards me.

- Given this postulate, we must create a compatible world model of this assumption, a model that authorizes application of Lorentz's transformations.

- With a bit of effort and imagination, we are able to create this model, which is described in chapters 1 and 2 of the 1905 paper, and which Einstein feels he has described in a complete and consistent way.

Applying this scheme, the sense of ED1905 becomes clear, and it is time to check if the overall reasoning is admissible. The last item of the scheme is the crucial one and is always inappropriately addressed first by the literature: authors try to make the reader understand first the reasoning on clock synchronization, and thus the reader is not properly informed of the premise constituted by the problem that had to be solved, and the effort needed to give a sense to the world model in which the speed of light is an absolute constant becomes immense.

Now, we might strive to better understand the world model where c is an absolute constant using the arguments found in relativistic literature, which are both few and many, according one's point of view: on the one hand they are few because they repeat always the same patterns of argumentation, on the other hand they are as numerous as relativistic literature is voluminous. But those arguments are all variants of what is described in Chapters 1 and 2 of the 1905 paper, and they add nothing to it unless the attempt to express more effectively the content of the theory. In particular, in paragraph 7 of Chapter 2 there are a few lines expressing all that can be said and was said to explain the kinematic consequent from the second postulate. Since the literal meaning of Einstein's reasoning is known to us by the careful reading of the crucial chapters of ED1905 we have made, we are able to move to the most important question of all, that well formulated breaks down into three alternatives: in the model of special relativity, do the clocks mark different times because somebody decides to synchronize them according to certain techniques, according to a convention? or do they not mark different times, but they just appear to mark different times? or do they really mark different times for natural reasons?

There are no other cases. The space and time dilatation can be interpreted as:

- *Conventional*: We have conventionally assumed that the speed of light is the same in every inertial frame, so we have to see what other conventions we must adopt to be consistent with this primary convention. We must wonder whether and why it is justified to apply Lorentz's transformations to what is not conventional, but physically actual, such as the existence of an electrical apparatus with its objective properties.

- *Apparent*: We believe that the speed of light is the same in every inertial system in an actual, not conventional sense. In this case dilatation is a kind of vision in perspective with which necessarily the inertial systems appear to each other, so also in this case we must wonder how Lorentz's transformations could be applicable to what is not apparent but physically actual, not to the light image of a thing but to the thing measured in its environment.

- *Real*: We believe that dilatation occurs as a natural process, such as the Earth's motion around the Sun or the vegetation of plants in the rainforest. Clocks mark different times for natural reasons, and in this case we have to wonder how the physical world is structured accordingly. Obviously in this case there is no doubt about the legitimacy of the application of Lorentz's transformations to the measurements of inertial systems in relative motion.

We shall now scrutinize these three cases by rigorously defining the meaning of the terms *conventional*, *apparent* and *real*, used so far having confidence that they have a generic but certainly clear sense for anyone.

If we find that only one of the three interpretations is admissible following logic, we shall have to choose that one. If, however, we find that two or all of the three possibilities are acceptable, we shall have to discuss which choice is the best, or arbitrarily choose one, unless we find no reason for choice. Finally, it may be that all three interpretations are logically untenable: in this case the problem would change in nature.

3. Einstein's solution

Of course, those who know and understands the theory of special relativity already know the answer and know what the correct interpretation is and why, but here we assume that there are many readers for whom the problem is not solved.

3.14 Conventional interpretation of dilatation

In this first case, the interpretation is: let us postulate in a conventional way that the speed of light is the same in every inertial frame. This would mean exactly: it is not known if the measurements always return c, but we remove the problems of measuring the speed of light, of the observer's speed with regard to the ether and of the choice between the theory of ether and that of emission, simply setting the speed of light equal to c with respect to all systems in relative motion, and adjusting accordingly each other measure.

In order to frame the meaning of the problem, we must first consider that speed by definition is the ratio of two fundamental quantities, space and time, and that if by convention we set a given speed as constant, then conventions with which to perform space and time measurements would always be necessary and would not cease to be in force: we would need space and time conventions anyway, to be able to measure lengths, durations, and speeds. Certain conventions allow us to have a fundamental space sample and to carry out measurements of lengths, other conventions allow us to have a time sample and to perform duration measurements. Given these conventions, all speeds are the result of space and time measurements, and not only do we not need to assume further conventions to have a sample speed, but it is not the case to stipulate them, because assuming conventionally a given speed as a constant sample we might find speed measurements contradicting space and time measurements. If given the measurement conventions and techniques that were adopted we say that a certain space Δs is 1000 and a certain time Δt is 10, it follows that the corresponding speed is $\Delta s/\Delta t = 1000/10 = 100$. But if for some reason we add the convention that precisely that speed is always 200 and not 100, we would therefore have to check and modify the space measurement conventions or those for time measurement to make the measured speed match to the conventional one. Since we have two fundamental quantities, space and time, to measure which is necessary to conclude conventions, there is in principle no reason to introduce other conventions regarding speed, which is not a fundamental magnitude, but is completely defined on the basis of the first two. However, given the complexity of the specific problem of the speed of light, let us suppose that it may make sense to decide conventionally that it is constant. But what previous conventions should we modify or what other conventions should we adopt to be consistent with the new convention on the speed of light?

Einstein's words at the beginning of the paper leave the possibility that the correct interpretation is this: "wir wollen die Voraussetzung einführen", or "we shall introduce the postulate" that the speed of light is an absolute constant. What Einstein in person thought to be should not be important, because what matters is to understand how the whole theory can be consistent even progressing further than Einstein, but the fact is that the theory of special relativity has not been subject to any modification after Einstein's original formulation, and this suggests listening to the implicit indications of Einstein's expression, which is closest to the original problems that the theory was born to solve. Now, in the text, the German word used is not "Postulat", which would mean "postulate" in the strong sense that the word has, but it is "Voraussetzung", which could be translated with the less demanding term "supposition". But above all, the text gave us a strong indication in the sense of conventional interpretation when it pointed out that a clock considered in motion would not delay by itself with respect to a clock considered at rest with which it had been synchronized, but should be tuned so as to mark the time of the moving system (a clock "adjusted so that it indicates the time t'").

Conventional interpretation is strongly connected with conventionalism, which was a current of the methodology of science that considered the focus of its interest the inquiry about what is conventional in science-building. Conventionalism carefully sought out the implicit and unexpressed assumptions that scientific thought postulates continually without realizing it does, and as a consequence of which many assert are exchanged for objective when actually they are arbitrary and conventional. Conventionalism is linked to the figure of Henri Poincaré, and it was a prevalent way of thinking at the very time when special relativity was born; Einstein was influenced by reading Poincaré's essays in a German translation just before the year 1905[28].

By analysing method problems through its principle, conventionalism has been successful in revealing the conventional character of concepts that in the past were considered natural, or otherwise unconventional. An irreversible success of conventionalism is precisely its concept of the clock: for how human thought is structured, capable of thinking relationships between things, but never finding absolute foundations, we have to give up the naive idea of looking for a clock to be considered absolutely valid. Instead, we define each clock as the counter of the same phenomenon, and we choose as the sample clock, the "right" clock, the counter of the repetitions of a periodic phenomenon, whose repetitions we consider all identical because given the current state of our scientific knowledge we have no reason to attribute them any difference. So what we do is to imagine that representing on a straight line the durations of the repetitions of our "right" clock, the segments placed next to each other have all strictly the same length, but this representation demonstrates nothing: it represents the meaning of our assumption, and it is not at all a check.

This attribution of identity to individual repetitions is of conceptual origin, not perceptional, although biological clocks, and therefore perceptional evaluations of time, play a role in the primitive construction of rudimentary clocks. In ancient times, when people built an hourglass and compared it with time from dawn to sunset, and chose the hourglass as a source of constant durations and recognized that the duration of the sunlight changes according to seasons, they were already informed by biological clocks and perception that the choice had to be for the hourglass: we know that the sun hours are different between summer and winter given the concomitant climatic difference and the great importance that this difference has for our body. But as soon as we pass over this elementary stage, the choice between clocks becomes conceptual and results from theoretical evaluations depending on the whole physical knowledge we possess. We chose the clock by which the whole of

---

[28] See Miller 2001, Chapter 6 p. 202 and passim.

experiences and cognitions we have receives a simpler and more consistent description. Thus, for example, at present the atomic caesium clock is considered the sample clock, and the quartz clock is considered inaccurate with respect to the caesium one. Obviously, no one can evaluate the constancy of the frequency of such machines with perceptional considerations: the evaluation that led to the adoption of the caesium clock as a standard is based on specialist considerations made by those who have the knowledge to be considered authoritative regarding the choice more appropriate, and for this reason, the choice in itself has an unavoidable arbitrary element beside rational considerations.

All conventional choices follow the criterion of simplicity and consistency, but the two criteria are not always applicable together. Sometimes simplicity suggests sacrificing consistency, sometimes the opposite happens. Thus, scientific conventions are eventually stipulated each time through a further convention, which decides on a case-by-case basis between simplicity and consistency: and therefore the work can never end. New scientific discoveries shed a different light on the conventions previously stipulated, and force us to change what was consolidated, endless.

By a simple, completely imaginary example, let us now see how the conflict between simplicity and consistency can arise, and how it needs to be resolved with a decision that is always in turn an arbitrary one. Let us suppose that all the events taking place on the Earth have a slowdown whose maximum occurs at the time of the winter solstice, and the minimum at the summer solstice, a slowdown we do not notice because it affects all on the Earth: all phenomena and all clocks are slower in winter and faster in summer. Let us suppose then that we have discovered with certainty that this phenomenon does not exist on Venus, and that the clocks installed on the Earth are delayed compared to those on Venus for part of the year, and for another part of the year they are expedited. If the data were just these, we would not have any reason to choose as the "right" clock one or the other: we could not know whether it is on the Earth or on Venus that clocks, such as caesium ones, give "right" sequences of repetitions, i.e. always equal, of the phenomenon chosen as the basis for the construction of the clock. The clocks on both planets could still disagree with a clock mounted on a third planet.

In this case, we would undoubtedly choose the Earth clock as a standard: if we chose Venus, we would then have to rectify the time measurements made on the Earth by introducing a huge practical complication not compensated by any advantage. In fact, choosing the clocks of Venus, and then constructing Earth clocks designed to mark the time of Venus, a thing which could also be feasible in the era of electronics, then in the Earth we would have clocks whose time would no longer correspond to that of the terrestrial phenomena in every season. The duration of cooking chicken would no longer be the same in every season, so we should also, in addition to the clocks, adjust the mechanism of rotisseries; the natural phenomena that we cannot control with technology would have discordant duration with respect to the seasons, and we should take it into account in every practical application.

The criterion of the choice would therefore be simplicity, and there would be no doubt about adopting this criterion, because we would not sacrifice any principle of coherence. But suppose we find out that the seasonal dilatation and contraction of the durations of Earth's phenomena depend on an interaction with the Earth's magnetic field. Since the planet Venus has no magnetic field (this does not belong to the example, so today is believed), we would now have a consistency reason to privilege the Venus clock: in fact, now we would know that the repetitions of the clock cycles on the Earth are different from each other for a precise cause we know, while the repetitions of the clock cycles on Venus would continue to seem identical to our judgment, because we would have no reason to consider them different from each other. Since the clock is chosen, by definition, choosing a phenomenon whose repetitions we all consider to be the same and not distinguishable from one another, coherence would make us choose the time of Venus as our standard; but instead, for simplicity, we would continue to choose earthly time, so we should avoid having to adjust every day even the rotisseries, the engines of every kind, and so on. Since the interactions among earthly phenomena we have to take into account are infinite, and the interactions of the Venus phenomena that are of interest to us are few or nothing and are of no practical relevance, simplicity would be preferred to coherence.

There is an actual example of this case: it is empirically demonstrated, without the need for relativity, that GPS caesium clocks accelerate by a small factor given the absence of gravity conditions in which they work in orbit. The discordance is small, but significant given the precision needed for the calculations of the position of the Earth receivers, that is, of the smartphones we all use. Of course, the practical choice is to correct satellites, and yet the reasons for the consistency might say: we should correct the Earth clocks because the frequency of the clocks working in the absence of gravity should be privileged as a standard, since there are different gravity situations for each planet and even at different latitudes on the Earth. The behaviour of the clocks where gravity is absent could suggest choosing a clock in that condition as the sample clock, instead of the terrestrial condition where gravity is not null and not uniform. No one would, however, make a similar choice, which would force us to adjust all the Earth clocks to agree with satellite clocks of the GPS system, rather than adjust the clocks of the twenty-four GPS satellites that are currently in orbit.

Criteria of simplicity and theoretical consistency can, however, come into conflict, so we know that tomorrow, when new natural properties of things unknown today are discovered, we may be forced to question the choices made and the conventions previously stipulated. The imaginary example of the discord between the clocks on the Earth and Venus shows us the impossibility of finding absolute criteria of choice: the final choice depends on the whole physical knowledge of a given historical moment, and also by inclinations that depend, so to speak, from the mood of every historical moment. There are no absolute criteria to know what choice to make except to always keep an eye on what we are choosing and the reasons for choosing it, and also the arbitrary component of our choice. And this general attitude is taught us by conventionalism as methodological theory.

Before conventionalism, one could believe that it was sensible to look for the "right" clock in the absolute sense; after conventionalism, each one understands that we cannot achieve such a goal, and this is a success of conventionalism. Conventionalism, however, after giving what it could give, then declined because the excess of conventionalism leads to the loss of the sense of reality, and to consider solved by conventions even what cannot be, and which requires patient empirical checks. On the extreme, conventionalism leads to say, "I'm sick, but instead of looking for the medicine that could heal me, I assume by convention to be healthy". From the point of view of incurably ill people this could be a wise choice, but it is not from the point of view of a patient for whom the cure exists. The clock case teaches it: careful research is needed to find the phenomenon (for example, the behaviour of the caesium atom) that makes the overall experience consistent and its choice most appropriate. After 1970, the caesium clock was chosen as the standard for determining time because, based on non-definitive considerations, the things we expect to happen at the same time seem to be measured more correctly or more comfortably by this kind of clock that others. For precision, before the caesium clock, the standard time was determined by the Earth's revolution around the Sun, hence on the basis of the solar year, and even before on the basis of the Earth's rotation on itself, hence on the basis of the mean solar day. The time of the caesium clock differs slightly from that marked by the solar year, and as we have seen above it is for this reason that every few years the international standard time is corrected by artificially adding a second. But the time of the caesium clock was chosen because, although slightly disadvantageous for astronomy, it proved to be more beneficial for air navigation control and other practical problems associated with recent technologies[29].

This digression was necessary to focus on the character of conventions, which necessarily intervene where there is no chance to find a rational criterion for a choice to be made. An immediate and naive attitude believes that we can always find rational foundations for choices and can always avoid the arbitrary element of the bet, and this is an illusion, but the necessity of conventions does not relieve us of the need to conclude conventions coherent among themselves: not every convention is equivalent to each other. That being said, let us suppose we want to take the conventional approach to the principle that c is constant in all inertial systems even in the absence

---

[29] See Essen Ray 2015.

of the ether. Assuming that the principle is conventional means that clocks considered moving are adjusted ("gerichtet", as Einstein says) to mark the calculated times t' for moving systems, when if they run undisturbed they would continue to mark times t to which they had been synchronized at rest. If that were not the case, the difference between moving and resting systems would not be conventional and the law of inertia would not be universal. To give visibility to the idea, let us also assume that the speed of light is lower than it is, and that Lorentz coefficients give rise to small but observable and significant values in some practically relevant circumstances. Suppose that on an airplane trip at 300 m/s for a few hours the Lorentz coefficients give rise to observable values, and for example we could have $t'=1.001 \times t$, which is a difference of one per thousand. We know that it is not so, and that given the value of c the values of Lorentz's transformations are much smaller than that, but this does not affect the reasoning in principle. What would happen? First of all, in this case, we should decide which clocks we should apply Lorentz coefficients to: to those of the aircrafts or of flight controllers on the Earth? We recall that the motion is relative given the law of inertia, so we would not have an objective reason to decide which clock to transform according to Lorentz and which to keep as reference. We would probably say: let us apply them to the planes, for simplicity. So a plane takes off, and pilots remember to adjust their clock so that it marks the Lorentz time t'. In doing so, we might say pleased: we have preserved the invariance of Maxwell's equations. This process could be automated due to the electronic technologies available today, but this is not the point. Once forced to mark the calculated time, the clock of the plane would disagree with that of the flight controllers, because the clocks, if they were undisturbed, would mark the same time. In our example, a 10 hour flight with a coefficient of 1.001 would give a 36 second difference between the two clocks. Then we should agree that even the flight controllers' clocks apply the Lorentz transformations. But then flight controllers' clocks would no more match the ones of aircrafts landed and at rest on the Earth, of railroad stations, and all the others. Moreover, the clocks of flying planes would no longer be transformed according to Lorentz compared to flight controllers' clocks: they would instead mark the same time. Ironically, in the end we would again have the problem of Maxwell's equations invariance, because after countless unnecessary adjustments all the clocks would go back to the mark the same time.

Then we could say: given the mess of practical problems, let us apply the convention only to electromagnetic problems. And so? We apply the criterion to the study of the behaviour of an electric motor in motion, and again we find ourselves to force clocks to mark discordant values with others. And then, today's clocks are all electronic, so we would be fully into the problem because we would not know how to act on the clock engine itself: should we apply to it the calculated time t', or not? I do not go on because the talk degrades into comedy: the primary occupation of everyone in the world would be to reconstruct the story of corrections made to the time marked by their clocks trying to reconcile the practical life requirements with the time measurement errors voluntarily introduced to preserve the form of Maxwell's equations. Since we have no "right" timepieces, but clocks designed to make coherent the whole of physical knowledge that we possess, forcing clocks to mark values differing from those marked by the optimum clocks that we have been able to build, would lead to the loss of the coherence of the physical world which we know, with a mess of unmatched empirical data and conflicts among experience data, and to live to create continuously new conventions to correct the consequences of the foregoing ones.

So, what could be the reason for correcting t in t', when the clocks mark both t? That of saving the form of Maxwell's equations in a fictive way, which may seem to be a result, if not useful, to be taken into consideration, only because of the complexity of the problem and the large number of elements involved, but it is a blunder. We would face an unmotivated convention in which the criterion of simplicity would be applied in an elephantine way to find a solution to the problem of electrodynamics at all costs, and would conflict with the criterion of coherence until devastating the coherence of all time measurements in the world.

The conventional interpretation of Einstein's postulate cannot be used, even though the origin of postulate has a kinship with the atmosphere of extreme conventionalism that was widespread in its time. One could still say: let us do nothing, let us not touch any clock, and get away saying that the convention of c constancy in all inertial systems saves the invariance of Maxwell equations. Which is to say: we have a problem, but if we conventionally assume that we have solved it, we shall solve it. Yet we shall see that this latter option, which could be called extreme conventionalism, had some credit at the beginning of the affair of relativity.

3.15   Apparent interpretation of dilatation

Einstein's paper belongs to the era of conventionalism, and the strongest interpretive suggestion we come up with at first glance is that of conventionalist interpretation. Instead, the strong suggestion that comes to us from relativistic literature after Einstein is to interpret the contraction or dilatation of time and space as an apparent phenomenon. The difference is not merely verbal: this case can be considered different from conventional interpretation by applying a precisely defined distinction criterion. In conventional interpretation, we stipulate that the speed of light is to be kept constant even if the measurements give different values, and we implicitly accept the need of corrections of the other measures to keep the accounts in balance. In apparent interpretation, instead, we say that lengths and durations are measured according to conventions that define measurement methods and time and space samples, but that light has the physical property to return the c value in any measurement, and that the consequence of this is that the measurements of lengths and durations of a system performed by another system moving with respect to it are necessarily different from the measurements performed on itself by the same system at rest. It is not easy to build a kinematic model in which this situation is conceivable outside the original hypothesis of contraction with respect to the unique frame of ether, but the sense of the words is univocal: it is assumed that anyone who measures the speed of light always gets the result c (even in the emission hypothesis and for a beam with respect to whose source the measuring device is in motion), and consequently it is assumed that those who perform remote measurements of a system in motion with respect to themselves would always detect measures different from those detected at rest in the measured system.

It is said, with a technical term constantly used by relativistic literature, that each system has its "proper" time in its frame, and its "proper" space in its own frame, and different times and spaces in different systems with respect to which it is in motion. The proper space and time are precisely the result of measurements at rest, conceptually analogous to the elemental operation of comparing the lengths with a ruler and the durations with a clock. The term "apparent" can therefore be understood as synonymous with the expression: each system has its own "proper" time and space, and has different times and spaces in the measurements made by the different systems with respect to which it is in motion: these will be called "apparent". This expression is recurring in relativistic literature, and if we use the term "apparent" to avoid repeating it, it is not necessary to understand the term "apparent" as opposed to "real". The apparent measurements are not different from the "real" ones, which if you want do not exist, but are remote measurements performed by electromagnetic signals, and are different from the measurements performed in the systems at rest using the conventions that have been stipulated for the local measures, not deformed by the constancy of the speed of light.

Apparent measurements are not only different from measurements at rest, but are manifold, while measurements at rest applying a given standard are unique and univocal. This happens in every case where we can distinguish between fundamental and unambiguous measures and apparent measures, and the situation is commonly met. To measure the height of an object once the standard of length is defined and has been agreed that the height is the value that is read when the distance between the object to be measured and the metric rod is zero, we must match the object to a graduated ruler. If we look at things in perspective, and put in front of us the metric rod and the object farthest, the object will look smaller than it appears when it is close to the metric rod. On the other hand, if the object is closer to us than the metric rod, then it looks higher. So let us say that the apparent measure is the one that disagrees with the standard procedure defined (by means of conventions) to return a unique measure, and that it can have infinite values in relation to some parameter, while the standard measure gives only one value.

This suggests that the experience of dilatation/contraction of space and even of time in a way corresponds to a trivial experience, though not in the sense of special relativity. If we see a man moving away, we see him becoming smaller and smaller: his apparent lengths contract, mutually we can say that his space dilates, and under certain conditions we can say that his time also dilates. In fact, if the man who is moving away has a clock that gives a sound signal every second, as the sound travels at a finite speed belonging to the speeds of mechanical world, 340 m/s, the interval with which we hear the sound signals would increase while the man walks away from us. We could then say that his time segments contract and that his time dilates until the man returns at rest with respect to us: then the sound signals would again reach us with the frequency of exactly a second, all with the same delay, and we could say that the time dilatation has ceased.

In the conventional interpretation of contraction/dilatation we had the problem: how to apply the convention? What clocks should we transform according to Lorentz? We have seen that the problem is insoluble: any choice would lead us to put clocks without reason in an error condition that would end up reproducing endlessly after each new adjustment. In the interpretation of the contraction/dilatation as apparent phenomena this problem does not arise: Lorentz's mathematics defines a kind of relationship between inertial systems which is similar to seeing in perspective: each inertial system sees the other systems under this deformation, whose equations are those of Lorentz, and there is no minimal logical difficulty. Two people go away, and each one becomes a dwarf in the eyes of the other. The rule is not exactly that of perspective view, because in the case of perspective when the two persons become again at rest being distant, they remain small one with respect to the other, while in the case of special relativity when they stop the space measurements become again equal to the ones measured at rest. In the case of time, the situation resembles more than that depicted by relativity: while a person gets farther away, the acoustic signals of his or her clock would be distanced by ever increasing times due to the increase in distance, but once the remote clock stops we would return to hear its signals with the same frequency with which we would perceive them locally.

Therefore, the difference in the measurements can be considered apparent without contradiction. Among systems in motion there is a relationship whereby the measures taken remotely differ from the univocal ones performed locally with standard techniques, and differ from the "proper" time and space of things, and have infinite values. One might object: but how and why should this happen? What process of measurement leads to measures discordant from "proper" time and space? The theory mentions, but does not develop (the complete development never occurs, even in the relativistic literature) the measurement procedures which, if implemented, would result in measures transformed according to Lorentz transformations with respect to the "proper" measures. We could try to integrate the theory by defining remote techniques and measuring procedures for moving systems, but we would immediately strike against an obstacle, which consists of the fact that the theory fully indicates the coefficient of transformation: so, applying it, we would always get "proper" measures. So why should we take the apparent measures when we know the rule to transform them into "proper" measures? Why should we take ambiguous measures on purpose when in any case we can know the unique measures? The difficulty is great: the theory speaks, but does not develop, the idea that reciprocal measures between inertial systems are necessarily transformed according to Lorentz, but this would happen only in two cases: it would happen if some people who do not know the theory implemented procedures which give rise to this deformation and ignored the correction factors because they do not know anything of it, or, strangely, it would happen if some other people, even knowing the theory, omitted on purpose to take it into account, exactly in order to ensure that the reciprocal measures of the systems in relative motion correspond to the predictions of the theory. The problem is that a distortion determined by the context of the measure may be necessary and inevitable for perception, but it cannot be necessary for the intellectual judgment of those who know the problem. If I see people shrinking in perspective because they are moving away, and I know the mechanism of vision, I cannot avoid seeing the persons becoming smaller as they walk, but I cannot be forced by any need to commit the error of measure I could commit if I were in a state of ignorance of the mechanism of vision and perspective. For example, imagine that a man two metres tall is far from us, and putting the ruler between us and him we measure his height in twenty centimetres: because we know the reason why in these conditions we measure twenty centimetres and not the two metres we would measure by putting the man in contact with the rod, what necessity could ever force us to mistakenly judge that man to be twenty centimetres high?

3. Einstein's solution

All measures that are not trivial, such as a tailor's, assume a theory of the object to be measured and require the development of calculations. At the time of Christopher Columbus, the accumulation of astronomical measures referred by travellers to Europe, Asia and Africa had allowed an exact enough map of the world known in latitude, but very uncertain in longitude attributes: Columbus, being convinced of certain misconceptions about the longitude of the Chinese coast, believed that a ship could reach it quickly crossing the Atlantic Ocean, and as we know the consequence was the discovery of America. The latitude attributes were more accurate than those of longitude because to know the latitude of a place is sufficient to take into account the calendar (the date compared to equinoxes and solstices) and the position of the Sun. To measure longitude, it is necessary to make observations and calculations that take into account both the Sun and the stars, or have a precise clock that marks the time of a reference meridian, such as Greenwich. Since the clock was completely missing in the Middle Ages, and it was even harder to take into account the stars in the observations rather than only the Sun, measuring longitude was more difficult for travellers than measuring latitude. But in both cases, the travellers who collected data had certainly not measured the Earth with a tape measure while travelling on the Silk Road or along the Mediterranean or African coast; the observations of the Sun's height at noon and of the position with respect to the stars taken by the travellers had accumulated for centuries, and a development of calculations had led to the formation of a world cartography as accurate as possible, with measurement errors more relevant to the longitude than to latitude.

The reciprocal measures of remote systems are in a similar situation: a theory is required to execute them, measurement errors can occur also for misconceptions in the theory, but if there is progress in the theory of the object to be measured, then the measurement errors are corrected and measures acquire a better quality. So it is impossible to understand why people should introduce a mistake that they know and which is described by the available theory. Therefore, it is reasonable to doubt that relativistic literature (which compared to Einstein's text is more likely to interpret dilatation as apparent) tells a completely sensible thing: in the hypothesis that the principle of the constancy of the speed of light in all inertial systems is valid, those who did not know that principle could make measurement errors as a result of their ignorance, but those who knew it would possess a method outlined by the same theory to correct those mistakes. The idea that a mistake must be accomplished on purpose or necessarily by those who know the way to avoid it is bizarre: it is true that people often behave like that in sentimental affairs and passions, but not in relation to technical issues.

But let us admit we can conceive in some way the case where the "proper" measures of a system necessarily differ from those taken by other systems in relative motion. The measures taken by other systems would then be apparent. The problem is: what is this distinction for? If a tailor took someone's measure looking at him or her in perspective with respect to the metric rod, instead of looking at him or her close to the rod, and then made a coat for him or her by measuring the fabric with the standard procedure, that is, by applying the fabric to the rod, what would happen? The tailor would have a coat that would either be too big or too small for those who should wear it. So even if the relative motion condition necessarily led to measures differing from the "proper" ones, what would we solve in relation to the problem of Maxwell's equations invariance? Nothing, because then in the case of an electromagnetic system considered in motion we would replace the standard measurements x, y, z, t with the measurements transformed according to Lorentz, where t is not equal to t', instead of with the measures transformed according to Galileo, where as we know $x' = x - vt$, and $t' = t$. But the measures that are transformed according to Galilean rules would be those that correspond to the "proper" and univocal dimensions of things, while the measurements transformed according to Lorentz would correspond to infinite apparent measures taken by observers in motion at all the infinite different speeds that are possible. If Lorentz's coefficients were significant in relation to technical practice, applying them we would cause damage similar to that made by the tailor of our example. The apparent interpretation shares with the conventional one the fate of not solving the problem of Maxwell's equations invariance, if not from the point of view of the mere mathematical form.

3. Einstein's solution

To sum up the above, let us ask two or three questions. Let us suppose that it makes sense to distinguish between "proper" and apparent time and between "proper" and apparent space. Let us first ask: to whom does the apparent measure appear? The theory has given us a perfectly precise and unambiguous answer to this question, and it answers in a doubtless way: apparent measurements are those resulting from calculations made by an observer measuring by electromagnetic waves another system in motion with respect to itself and developing calculations by applying the rules that should apply in the case of emission theory, adding the speed of the waves to the speed of their source, and thus ignoring the fact that c is a postulated constant for every observer at rest or in motion. As we have seen above, it is paragraph 7 of §2 of ED1905 that provides this answer: the measure is not the same as that made by the assumption of the theory of the emission. But even if there were no such explicit words in the original Einstein's paper, literature, and all the theory do not allow an answer other than in this way. Now, however, we have to ask ourselves a second question: why should an observer who has studied relativity then take measures by neglecting to apply Lorentz's factor in order to convert the dilated measures into the "proper" ones? And finding an answer is a hassle: why should one make a mistake when one knows the rule to avoid it, if not to create the illusion to him or herself that his or her sweetheart is faithful, or in other similar situations? And finally, a third question: if anybody did take measures that according to the theory of relativity are incorrect, what consequences would this have for the physical world outside human judgement? What physical properties of the world would be revealed? What importance does this event have? None, and therefore apparent interpretation, as well as having little or no meaning in general, has no utility, not even in relation to the problem of Maxwell's equations invariance.

If all this is not clear enough, let us see a rudimentary example, not exactly corresponding to the relativity problem, but analogous and perfectly suitable to illustrate in the simplest way why conventional or apparent interpretation cannot solve a problem of the kind of Maxwell's invariance. Let us repeat the example which we have just mentioned above about two men A and B each having a clock that emits a beep every second. One of the two, let us say B, sets off and heads forth at 34 m/s. As sound propagates at a speed of 340 m/s, both of them hear the beeps of the other with a growing delay, and each one judges that the clock of the other has slowed by 10%. After 10 seconds B stops and A sees that his clock marks 10 seconds past the beginning of B's trip, but he knows he has counted 9 beeps from B, so he judges that B's clock is delayed by one second. Instead B's clock marks 10 seconds and emitted 10 beeps, but the last beep starts when B has travelled 340 metres away, and will come with a delay of one second when A's clock will mark 11, and from now on B will have a constant delay of one second and no longer an increasing delay, unless he begins to move again. At this point, taking into account the delay of beeps due to the finite sound speed, both A and B can correct their remote measure, but they will have no reason to attribute the apparent time they had calculated before the correction. If they did, they would introduce a mistake: A would say that B's clock marks 9 seconds when B's clock marks 10 as his one, a fact that A knows well if he understood the reason for the apparent slowdown of B's clock while in motion. Instead, if A believed that sound propagation is instantaneous, as a consequence of this error he would believe that B's clock lags as a consequence and a function of the motion condition, but that would not change the physical state of clock B.

With Maxwell's equations, the case is the same, although the situation is more complex: considering the contraction of durations apparent, whoever applies the transformations to the state of the remote system would introduce an error instead of correcting it. And yet, our discourse has become paradoxical because it is impossible to accurately analyse the consequences of the case where dilatation is apparent, since the very concept of apparent dilatation is impossible: a theory cannot consider as necessary a mistake that the same theory teaches how to correct. If Lorentz's coefficients had observable values, then as a consequence of the fact that the ether does not exist, Maxwell's equations in the form in which they are would give rise to observable errors, and it would be necessary to correct them not with conventional artifices, but by studying the problem with actual experimental and theoretical processes.

3.16 Realist interpretation of dilatation

Given the difficulties inherent in conventional and apparent interpretations, we need to consider the third case: the interpretation that contraction/dilatation would occur as a result of a natural process, and two clocks that in relative rest would indicate the same time, in motion would indicate different times just for being in motion. The durations of all physical processes would be altered in the same proportion as the clock movement, and people would have different ages in relation to the motion condition. The same would happen to spatial lengths, and those who would move at high speeds would be dwarfs with respect to those who are at rest.

Whether this is the interpretation to be given, it is still doubtful in the letter of the original text, that when touching this problem speaks of an "adjusted" clock, *gerichtet*, in order to mark the transformed time t' instead of time t. But even if we consider obsolete Einstein's original text and we turn to the latest relativistic literature, things do not change because the problem is discussed in detail by always assuming and using the distinction between "proper" time of a system and time measured in other systems. In a recent textbook we read[30]:

> Let us call *proper time* $\Delta\tau$ the time interval between two events when measured in the references system where the events happen ...

and the same notion is found everywhere because it is considered fundamental to define the notion of contraction/dilatation. But if there are one proper time and many non-proper times, the interpretation to be given seems to be the apparent one: if a clock is synchronized with another at rest, and then leaves for a journey, then from the point of view of the system left at rest while travelling it appears to mark a different time than the clock left at rest, but when it returns to the base it must mark the same time as the clock at rest.

For the moment, let us not ask ourselves how the theory solves this problem. Let us not wonder whether the theory in Einstein's expression or today's expressions supports apparent interpretation or realist interpretation. Let us not ask for now whether Einstein and the

------

[30] Ferraro 2007, p. 48.

next literature have given the same interpretation, or have given two different ones. Let us just ask if we could adopt the realist interpretation.

First of all, let us summarize what we know and formulate the problem without ambiguity. We know that in the hypothesis of the ether the speed of a light beam is relative to the ether, and adding Lorentz's contraction hypothesis (just for space is enough) has the consequence that the measured speed of a light beam is the same for all the observers. Then we know that Einstein's relativity has the purpose of abolishing the ether, so it is forced to conceive the motion of electromagnetic waves in the frame of their source, and we have seen that Einstein's argument takes this interpretation for granted. Then we know that the next hypothesis is that the measurement of the speed of electromagnetic waves in the vacuum is the same for all the observers either at rest or in motion with respect to the source of the waves. At this point, if we want this to be possible, the dimensions of the observers in motion with respect to the light source should be contracted according to the speed with respect to the source of the waves. The situation is the same as that described by Lorentz, except that the reference for speed is not the absolute frame of the ether, but the relative frame given by any source of electromagnetic waves.

Is this state of affairs conceivable? Let us suppose it is, so let us assume that two clocks in motion really disagree. At the beginning of a trip, two clocks are synchronized locally. Then one of them starts, later he comes back, and at the end of the journey the clocks mark different times, and this can be verified again with direct observation. Everything in motion has evolved in accord with the clock in motion, so everything in motion has aged less than things at rest.

Having made this hypothesis we have a first problem: the law of inertia and the relativity principle both say that the state of uniform rectilinear motion cannot be distinguished from the state of rest. We owe this notion to Galileo and Newton, and the intent of the theory of special relativity is, as we know, to preserve this principle by extending it to all physics. So, having to preserve the law of inertia, how do we choose which clock will slow down between two clocks in motion? In the empty space, two satellites travel in the opposite direction. As they meet the two clocks mark the same time. Later, which of the two will suffer the contraction of its durations and the dilatation of its time? Will it be the one I see go to the right, or the other I see go to the left? On the highway I see two cars travelling in the opposite direction at 100 km/h. The relative speed between them is then 200 km/h. Which of them is affected by the dilatation corresponding to the Lorentz values that are calculated for that speed? Here instinct answers: but it is clear that both cars undergo a contraction corresponding to 100 km/h because they are moving and the land is stationary. No! The theory does not say this. The old law of inertia, which Einstein wants to generalize and not to abolish, says that straight-line motion at constant speed is always relative, so that we can consider at rest either the Earth, or one of the two cars or the other. All three points of view are equally legitimate. So who contracts? Car A, or Car B, or both, or the whole Earth together with one of the two cars, because the other is considered at rest?

If trying to answer, you said "the cars move, and the Earth is still", committing an instinctive mistake you have hit the target of the problem: the hypothesis of dilatation of spaces and times is logically incompatible with the law of inertia. In order to know where time is dilating, and where it does not dilate, it is imperative to have an absolute reference system. For this reason, there is no difficulty in understanding Lorentz's hypothesis, where the absolute reference system is the ether, and therefore it is unambiguously defined who are those who suffer (given the hypothesis) the contraction of the lengths and durations and the dilatation of space and time: it is whoever moves with respect to the ether that undergoes the contraction. In that hypothesis everything is clear and unambiguous, and the hypothesis is conceivable. So when the readers say "the cars move, and the Earth is firm", when making a mistake they simply follow the instinct that leads them to reason properly: to apply the notion of contraction/dilatation an absolute and common reference to all systems in relative motion to each other is necessary.

A first, insurmountable problem is therefore this: the law of inertia does not allow to decide, between two systems in motion, which will be subject to contraction and which will not. It would be easy to conclude that the realist interpretation is impossible.

But there is a second problem. Just for the law of inertia, each system at every moment is in motion with respect to infinite other systems, and at infinite different speeds. The reader of this book is now at rest with respect to the Earth, or if you are reading while travelling you are at rest with respect to your form of transportation. But at the same moment, the reader of these lines is in motion at countless different speeds with respect to all vehicles and all animals that move on the Earth, with respect to the wind blowing in so many parts of the world, with respect to the Moon, the Sun, the planets of the solar system and all the stars. Which of these speeds is to be preferred to determine the Lorentz coefficients to be applied? Or do we mean that every inertial system, at the same present instant, is now experiencing not one, but infinite actual and effective contractions with respect to infinite other inertial systems?

Let us consider three vehicles as in this figure:

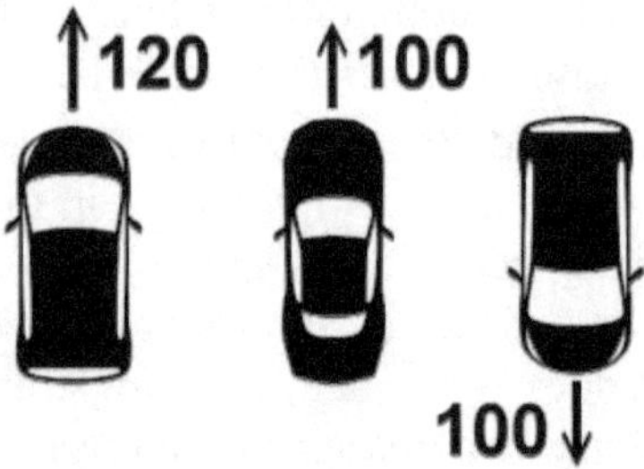

Let us suppose that as the three vehicles cross over, their clocks are synchronized with absolute precision. If their time must dilate according to speed and their clocks are delayed accordingly, which of the many relative speeds in the game will result in dilatation?

Considering two systems, it is impossible to decide which to attribute the dilatation, and with greater reason the decision is impossible if we consider that systems in motion are never just two as in the examples described in books, but are always countless. If we want to assert that a clock is delayed according to its speed, we implicitly have to assert that our clock at this time retards according to a given factor compared to the clock of those who move on foot away from us, according to other factors compared to clocks travelling on trains and aircraft, to other different factors with respect to clocks installed on artificial satellites, and so on. Under Lorentz's hypothesis, all these clocks would have some delay given the speed with respect to the ether, and in theory one could calculate the relative delay between any couple of them. But removing the common frame of the ether, what could be the reference speed? Our clock is one, and it cannot delay according to infinite factors except in a conventional or apparent way.

At this point, the readers' thought will go spontaneously to reconsider conventional or apparent interpretations, because the hypothesis of real contraction/dilatation is radically incompatible with the law of inertia. To keep together the law of inertia and the hypothesis of temporal dilatation we could try to hypothesize some distinction criteria of any kind with which to decide uniquely who is contracted, but then in applying the hypothetical criterion we would always have infinite possibilities excluding an unambiguous solution. For example, to have any criterion, we could assume that the system which contracts is always the one with the least specific weight. So we would have a unique criterion for establishing a contraction hierarchy. But let us try to see the consequences, and now imagine finding an object light enough, say a piece of wood, that is in motion with respect to a lead bar, and is at rest with respect to another identical lead bar. Since the specific weight is smaller, wood should be contracted and not lead, but as the lead bars are one in motion and one at rest, will the wood contract, or not? And if the two lead bars move both at different speed, what should be considered? And so it goes on, endlessly.

The unambiguous criterion could be only absolute motion, which is therefore a common reference system for all motions, even other than the ether, which would allow a determination of absolute speed. And so a consequence emerges that is very important and is also involuntarily ironic. Before Fitzgerald's and Lorentz's hypotheses, no one had ever thought of dilatation of time. Then the hypothesis of physical contraction with respect to the ether had made perfectly conceivable the dilatation of time, the idea of local times, and the case in which simultaneity gets out of phase and becomes succession. But then the general extension of the law of inertia proposed by Einstein returns to make the time dilatation impossible. Ironically, just the attempt of Einstein restores the unity of time. Not the unity of absolute time, but the logical requirement of the unity of time in which the human intellect places the phenomena. The course of thinking of the readers who study relativity and want to understand it, always tries to find a way to conceive the time dilatation according to relative speed, and hitting against failure immediately withdraws to the apparent interpretation. The apparent interpretation is problematic for other reasons, and so one comes back to the doubt that the interpretation should be the realist one, but what matters is that thought, living and reproducing this oscillation between the imaginable interpretations, realizes the inability of man to know anything that cannot be represented in the unity of a single temporal frame. Given a hypothesis such as that of Lorentz, the duration of the phenomena can be conceived as parametric, and the local times of the moving systems may differ, but the form of the Lorentz formulas shows that the durations are parametric compared to the unique reference of a fundamental time, the one at rest with respect to the ether. By generalizing the law of inertia, fundamental time becomes the purely logical and formal one of the human mind: and so thanks to the law of inertia this logical unity always prevails on every attempt to abolish it.

3.17   Balance of the three interpretations

Let us now check the balance of the three possible interpretations. Conventional interpretation is a bad convention. We know that space and time measurements are made on the basis of the assumption of time and space samples that make the overall experience as consistent as possible with the whole of the physical knowledge we possess at a given moment, and we know that these assumptions are destined to evolve. In the conventional interpretation of Einstein's relativity we would adopt as a further conventional sample a speed, that of light, when speed is a quantity already defined and measured on the basis of time and space samples, and therefore does not need a sample of its own. But the operation would be not only redundant, but also harmful: we have seen that if the supplementary convention were to be applied technically, we would create the problem of adding infinite further conventions for infinite clock synchronization problems, and these problems would exist only because this redundant convention about speed was adopted. The price to pay for the invariance of Maxwell's laws would be immense and the solution would remain fictitious, when in the end the subject of the interest of those engaged in the search should be the knowledge of the propagation mechanism of electromagnetic waves, if they do not propagate in the ether as scientists of the nineteenth century used to think.

The interpretation that space and time dilatation would be apparent gives rise to more than one problem. First of all, this interpretation would have no physical meaning in the same way the conventional interpretation does not have. Then, it is difficult to conceive a procedure in which appearance is necessarily determined by mediated and non-perceptional measurement judgments. It is like conceiving a situation in which some people know that they are seeing something small because they see it distantly, but although they know the criterion of the calculation needed to bring the measure to a standard value they do not apply it, because they are obliged by some reason to refrain from applying it. The theory, on the other hand, prescribes the method for correcting apparent measurements through Lorentz's formulas, and "proper" measures can always be calculated.

3. Einstein's solution

Finally, as apparent interpretation has abandoned the hypothesis of the ether, in this picture one continues not to understand how the speed of light can be independent of that of the emitting body. In conventional interpretation, the postulate of the constancy of the speed of light is a conscious arbitrary choice, to be discarded because it is inappropriate, but not inconceivable. In apparent interpretation, dilatation is the apparent product of unequal measurements, but the assumption of absolute constancy of c is interpreted as a real fact: hypothetical and ad hoc, such as the contraction of FitzGerald and Lorentz, but having physical meaning. But then, in order for the speed of light to be physically and not conventionally the same in the different inertial systems, the inertial systems should adapt themselves to this by a physical process of some kind, by contracting the measurement instruments with which they determine the speed of light, and hence the apparent interpretation evaporates by itself and prepares the road for the realist one. If the speed of light is really the same in all inertial systems, and therefore is never $c+v$, but always and only c, then inertial systems must adapt themselves to this for a physical reason: Lorentz's hypothesis asserts exactly this and explained how this could happen with a coherent kinematic, of which we did not find any trace in relativity.

The realist interpretation necessarily implies assuming an absolute reference system for motion. In fact, given the law of inertia, considering two or more inertial systems in relative motion, we would always find ourselves unable to decide which of them is contracted or dilated and which is not. So we should find a criterion for deciding. But this criterion could only be an absolute frame for motion, and this would imply abandoning the law of inertia, because with any other kind of criterion it would happen that every system would endure countless dilatations at the same instant, because each system is in relative motion to infinite others at different speeds.

Unless there is a fourth yet different solution, we cannot adopt the theory of special relativity to solve the problem of Maxwell's equations invariance. More specifically, what we know so far is that we cannot adopt any of the three interpretations we considered. To see if there is a fourth point of view, Einstein's original text does not help us anymore, and we shall have to see how relativistic literature proceeds. Before this, however, we must consider two more Einstein results, the undoubted success of the notion of electromagnetic quantum and the famous equation that establishes the criterion of quantitative equivalence between energy and mass.

4. *The relativistic concept of mass: $E=mc^2$*

A premise: this chapter is intended for both the reader who has no idea of the meaning of the famous formula, and the reader who understands its meaning, but does not know by what process it could be derived. Our sole purpose is to show how the equivalence between mass and energy can be conceivable. So we shall use didactic and historical materials that are not up-to-date and that we only need to lead us into a condition starting from which anyone then could deepen their understanding of the problem, if desired. I insist: this chapter is not intended to understand how mass and energy equivalence can be conceived in the era following the development of nuclear physics and in 21st century, but only to get an idea of that equivalence and its relationship with special relativity and with the postulate of the absolute constancy of the speed of light. On this latter issue we shall have to return even further, because in this chapter we limit ourselves to introducing mass and energy equivalence without using the concepts of special relativity to deduce it. We shall then see that in relativistic literature that equivalence is inferred from the specific relativity postulates, and that there are significant logical problems associated with the form of mass and energy equivalence.

Those who do not have the elementary physical school-level skills that are needed to read this chapter, if they cannot follow it, can go further: they will continue reading just what concerns relativistic kinematics and the problem of space and time measurements.

4.1   *Concept of photon, or electromagnetic quantum*

First of all, we need very few words to describe the concept of light quantum, which Einstein exposed in the paper "On a heuristic point of view concerning the production and transformation of light"[31], published just before the paper on electrodynamics.

Experiments had been made, and were underway, which had revealed an unexpected phenomenon. Technology of that era made possible to build instruments capable of emitting electromagnetic waves of different frequency determined precisely, and to construct tools capable of detecting observable and measurable effects of electromagnetic waves. It was observed that a given species of effects — I do not say which, because we need not accumulate unexplained notions, but only know the indispensable premise of our discourse — did not vary continuously when the frequency of the waves was varied, but only when a certain value of threshold was reached. Then the effect did not increase until a new threshold was reached, and then the effect increased again by leaps. There was therefore in input a phenomenon that was constantly changing — waves of increasing or decreasing frequency with variations as small as desired — and in output a phenomenon determined by the former that manifested neither in a continuous nor in a random way, but jumping in certain moments from one state to another discontinuously, according to a precise rule. This is the photoelectric effect[32]. In the contemporary interpretation of that effect, there was a difficulty because it was asked: how can light, or rather, an electromagnetic wave, which can vary continuously in frequency, produce an effect that appears in a discontinuous way? And physicists tried to formulate a response that would explain this transition from continuous to discontinuous. Einstein's paper, on the other hand, proposed a description of electromagnetic waves that interpreted them no longer as a wave similar to water, but as a sequence of indivisible elementary entities, whose behaviour is

---

[31] Einstein, Photoelectric effect 1905.

[32] Who needs a very clear explanation about this can find it in Breithaupt 2002, Chapter 8.

similar to that of a wave, and yet emitted in a discontinuous way. This characteristic of light explained the discontinuity observed in phenomena by attributing the character of the discontinuity to the same light wave, that is, to the cause of the phenomenon, rather than to some intermediate mechanism between the wave and the effect. These indivisible constituents of the wave are the so-called electromagnetic quanta, later called photons, and their definition, then evolved but no longer questioned, is why Einstein was awarded the 1921 Nobel Prize. The 1905 paper on the photoelectric effect reached Max Planck, who was director of the most prestigious German physics magazine, was published, and had a certain circulation among physicists who knew the problem.

That is all we need to know about electromagnetic quanta: they are indivisible constituents of electromagnetic waves that manifest themselves in a discontinuous way. It should be noted that Einstein's contribution did not consist of the theoretical elaboration of experiments performed by himself, but in drawing simple conclusions from known experiments that others interpreted in a complicated way to preserve the consolidated model, trying to maintain the continuity of light phenomena instead of accepting the fundamental discontinuity that they show to have.

4.2   *Basics of elementary physics needed to understand $E=mc^2$*

We saw that understanding the sense of special relativity theory about the contraction/dilatation of time and space requires after all just the cognition of what is speed. This does not immediately lead us to understand how this phenomenon is possible, and so far we have not been able to choose among the three possible interpretations we could isolate, but the notions needed to frame the problem are in fact few. What is problematic is the mechanism of deriving Einstein's consequences from his premises.

To get an idea of the way in which the mass and energy relationship expressed by the formula $E=mc^2$ is reached, is a simpler matter than one would expect, but it is necessary to review some classical mechanics notions and if necessary read them in some elementary or high-school textbook[33].

Here is a list of what you need to know in order to read the following. Those who do not know these basics must read something about them, or move to the next chapter, in this case just remembering that the formula $E=mc^2$ is not exactly a consequence of relativity, even if it is connected to it for several reasons.

The basic quantities of mechanics are three: space, time, and mass. What are these things? They are what everyone knows and nobody can define: the space around us, the time in whose frame we get old, and the solid, liquid, or gaseous things that constitutes our bodies and everything else, and that exists moving in space and changing over time. More cannot be said, and it is not necessary. But it is important to add that mechanics also uses the concepts of speed, acceleration, force, work or energy, and power, and that these concepts are not fundamental, but are defined on the basis of space, time and mass. Each of these concepts has its specific unit of measurement, and the situation is the following.

Space is measured in metres, and time is measured in seconds; we have seen that it is necessary to stipulate motivated conventions to define the metre and the time second in a useful way.

---

[33] For really elementary basics see Breithaupt 2002.

4. The relativistic concept of mass: E=mc²

Mass is measured in kilograms, and here conventions are agreed also to define the method for knowing what a kilogram is, and how to determine that any mass corresponds to a given quantity of kilograms.

Speed is the ratio between a space path and the time it takes to travel it, so it is measured in metres per second, m/s; Acceleration is the ratio between a variation of speed and the time needed for it to occur, and then we measure it in (metres/seconds)/(seconds needed for variation), that is, m/s^2.

Force requires some attention, because at first sight it would seem that it cannot be defined through the three fundamental quantities. But for the law of inertia, a mass maintains its state of rest or uniform straight motion until a force intervenes to modify this state, resulting in a variation of speed, that is, an acceleration. So, even if we do not know what force is from a qualitative point of view, the same way we do not know about the other quantities, we know that a force is something capable of accelerating a mass, altering its state of inertia. So a force is an acceleration impressed to a mass. The unit of measure is the Newton, which is the acceleration of one m/s^2 impressed to a mass of one kilogram: $N=kg \times m/s^2$. Force is therefore the ability to accelerate a mass, for example, to impress the acceleration of 1 m/s^2 to a mass of one kilogram, and hence, for example, the ability to lift a weight. Work then is the actual use of force by accelerating a mass, and hence the force multiplied for the space travelled by the mass to which force is applied, with any time duration of the process. Work is measured in Joule, J, where a Joule is one Newton × one metre, J=Nm. Work is also called energy; the two words are synonymous. Energy is present in different aspects of experience, but the reality that corresponds to it is the same: in particular, kinetic energy and heat are the same thing that appears in a different form. Power is the ability to make a certain amount of work in a given time. It is measured in Watt, where a Watt is the work of one Newton accomplished in one second, W=N/s. And finally, always recur the concepts of momentum, which is the product of the speed × mass and whose symbol is usually p, and kinetic energy, which is equal to $1/2mv^2$.

4.3   Convertibility between mass and energy

"A philosopher was asked: what is the weight of smoke? He said, weigh the wood before burning it and then the remaining ashes, subtract and you will know exactly the weight of the smoke". The example is by Kant, contemporary of Lavoisier: we all know that chemistry since the eighteenth century assumes the principle of Lavoisier, so that mass appears in chemical compounds under multiple forms, solid, liquid or gaseous, but in transformations it always keeps its quantity. In the same way energy is conserved: when two billiard balls collide, kinetic energy passes from one to the other, but the sum of the energy of the two balls remains constant. When a weight stops falling to the Earth, the lost kinetic energy converts into heat and raises the temperature of the site of the collision, but the sum of kinetic energy and thermal energy is conserved, and so on.

Nuclear reactions contradict this: the remaining ash and smoke (nuclear waste) weighs less than wood (the uranium used in the reaction), and the mass difference has been converted into enormous amounts of heat, that is, energy. The quantitative relationship between mass and energy is that expressed by the famous formula $E=mc^2$: if in a chemical reaction the mass of what is left after the reaction is less than the mass of what existed before, there is a development of heat, measured in Joule, equal to the lost mass multiplied by the square of the speed of light, i.e. by 9 with 16 zeros. We can say that for one kilogram of lost mass comes out a number of Joule equal to about 1 with 17 zeros.

The heat of a Joule is very small, because combustion of a match equals about 1000 Joule; but the loss of one kilogram of mass in a nuclear reaction develops the heat equivalent to that of burning a number of matches with 14 zeros: one hundred thousand billion. And if we knew a procedure applicable to the case, we would need 100,000 billion matches to produce one kilogram of mass of any chemical substance. Important: uranium and other particular metals are used in nuclear reactions because for such metals we know how to trigger and control the phenomenon of mass transformation into energy, but for any chemical element or compound the relationship is valid and theoretically it could be applied.

4. The relativistic concept of mass: E=mc²

The idea of the reciprocal convertibility between energy and mass requires that we stop thinking about energy and mass as distinct things: in the same way mass appears in a solid, liquid, or gaseous state, so the reality that exists in space and time and that presents itself to our experience appears both in form of energy and in form of mass: mass is "condensed" energy, so to speak, and energy is "melted" mass. What is conserved in chemical reactions is the amount of the mass and energy complex, not the mass alone and not the energy alone, and the principle of conservation of the mass that is applied in chemistry is approximate, not exact. The same applies to the principle of energy conservation. For example, when we set a mousetrap, by cocking the spring we lose energy and store it elsewhere: a cocked mousetrap is heavier than one that is not cocked. If to load the spring we use 1000 Joule, the energy of a match, then the loaded trap will weigh 0.00000000000001 kilos more than when it was not cocked. In nuclear reactions this phenomenon has observable dimensions, but it is present in every chemical transformation with tiny and unobservable values.

To imagine that mass and energy can be two ways in which the same thing presents itself is not difficult, if we apply the analogy with the fact that water and everything else can be in a solid, liquid, or gaseous state: ceasing to present itself as mass, everything can appear in the form of energy, and it is not difficult to conceive that that between mass and energy there is a transformation, as is that of ice into water or vice versa.

But how can the quantitative relationship expressed by the famous formula $E=mc^2$ be derived?

Experiments from the late nineteenth century were on the way to this relationship, which was intuited by Thomson, the discoverer of the electron, already in 1881, and by other physicists, including Einstein, who does not exactly have the paternity of this discovery. An online *Scientific American* article gives interesting data about its history[34], and there are numerous stories about on the Internet, that

---

[34] The title is: "Was Einstein the First to Invent E=mc²?".
https://www.scientificamerican.com/article/was-einstein-the-first-to-invent-and-mc2/

are to be taken with caution. In fact, many pseudo-scientific sites use the fact that Einstein was not the only one to perceive that relationship to demonstrate the most extravagant theories. It is true, however, that the discovery of the relationship is not to be attributed to Einstein alone. For curiosity, the quantitative relationship between mass and energy expressed exactly in the form $E=mc^2$ was also perceived by an Italian amateur, whose book looks like a maverick's fantasy work, but dates back to 1903, and thus shows how the idea was in the air in those years[35].

The relationship $E=mc^2$ is not necessarily connected to special relativity. Einstein put it in a three-page note[36] as a consequence of the ideas discussed in ED1905, but the connection with special relativity, or with the principle of absolute constancy of c, is not indispensable to infer the relationship. Indeed, later we shall see how this connection is the source of more than a serious problem.

Now we shall see how the $E=mc^2$ relationship can be derived from very simple premises and an elementary development, not connected to the theory of relativity, and having only an illustrative value, and then how Gilbert Lewis obtained it in 1908, expressly choosing not to hold account of the theory of relativity, which he also knew. Later we shall see the problem of connecting this formula to the theory of special relativity.

---

[35] The Italian amateur's name was Olinto De Pretto, and his research, published in 1921, can be found in the site www.liberliber.it

[36] It is the paper "Ist die Trägheit eines Körpers von seinem Energieinhalt abhängig?", *Annalen der Physik* 18, 639–641 ("Does the inertia of a body depend upon its energy content?")

4.4 *E=mc²: non-relativistic derivation by Rohrlich (didactic)*

The Austrian physicist Fritz Rohrlich[37] conceived a derivation of formula $E=mc^2$ in 1990 in a very simple way using only the laws of classical physics and the notion of electromagnetic quantum, or photon. This derivation is a simplification for didactic use. We cannot express here all the simple assumptions necessary to follow the argument: as already mentioned, only basic elementary physical knowledge is needed, so that those who do not possess it and want to follow the argument in every detail must obtain it from an adequate source. Otherwise, the reader should only observe the premises used by Rohrlich, because they are relevant to comparing the logic of the non-relativistic deduction of the formula with the logic of the relativistic deductions, which we shall see later.

The problem to be resolved is thus formulated: electromagnetic waves carry energy (trivially, sunlight warms). The mass-energy complex is constant and given this hypothesis a body that emits electromagnetic waves without being fed by an energy source, such as the Sun and the stars, must lose mass, i.e. convert its mass into the energy of light. We must find the quantitative relationship between mass and energy.

The premises used by Rohrlich are:

- the principle of conservation of momentum;
- the Doppler effect, for which by going towards a wave we increase its frequency, because the crests that invest in us in a time unit are more than those that would invest us if we were at rest, and by moving away from a wave the frequency decreases for the opposite reason;
- the constancy of the speed of light with respect to the emitting body, so without considering it absolute in the sense of Einstein;
- the notion of electromagnetic quantum and the law of Planck for which the energy of a photon is:

$$E = hf$$

---

[37] See Rohrlich 1990.

where E is energy, f is the frequency and h is a constant, that is, the Planck constant. The meaning of this formula is: light transports energy, as everybody knows after taking a sunbath; the light propagates in tiny indivisible quantities that have the properties of a wave, quanta, and the energy of each quantum is equal to the frequency of the wave multiplied by a certain constant. This law is derived from the interpretation of experimental data.

The process of the derivation of $E=mc^2$ is this: consider a material body B having mass m_1 moving with respect to an observer M with a constant speed v_1 very small compared to that of light. Also consider a second observer Q at rest with respect to B. Suppose that at a given instant t the body B emits two electromagnetic quanta having the same energy hf one in the direction of motion, the other in the opposite direction, where f is the frequency of photons observed by Q at rest with respect to B. The observer M then, taking into account the Doppler effect, will measure a frequency equal to

$$f\left(1+\frac{v_1}{c}\right)$$

where c is the speed of light in the vacuum for the photon emitted in the sense of motion, and will measure a frequency equal to

$$f\left(1-\frac{v_1}{c}\right)$$

for that emitted in the opposite sense. The radiant energy E emitted at instant t which is observed by M will therefore be equal to:

$$E = hf\left(1+\frac{v_1}{c}\right) + hf\left(1-\frac{v_1}{c}\right)$$

from which by simplifications is derived:

$$E = 2hf \quad (1)$$

For the principle of conservation of momentum, the amount of momentum of the body B observed by M prior to the emission must be equal to the sum of the momenta of B and of the two photons after the emission. The momentum of the second photon that is emitted in the opposite sense to the motion must be taken with the negative sign, so we have:

$$m_1 v_1 = m_2 v_2 + \left(\frac{hf}{c}\right)\left(1+\frac{v_1}{c}\right) - \left(\frac{hf}{c}\right)\left(1-\frac{v_1}{c}\right) \quad (2)$$

where:

m_1=mass of the body B before emission,

v_1=speed of the body B before emission,

m_2=mass of the body B after emission,

v_2=speed of the body B after emission,

$\left(\frac{hf}{c}\right)\left(1+\frac{v_1}{c}\right)$ = momentum of the photon emitted in the sense of motion,

$\left(\frac{hf}{c}\right)\left(1-\frac{v_1}{c}\right)$ = momentum of the photon emitted in the sense opposite to motion.

Given the symmetrical nature of the effect, after the emission of the two photons the observer Q will not detect any change in motion of body B, which will then continue to be at rest with respect to him or her. So for the observer M after the emission both the observer Q and the body B will continue to move at unchanged speed. It is therefore concluded that v_1=v_2. By substituting v_1 with v_2 in (2) and introducing the reduction of the mass m of body B after emission, equal to: m=m_1−m_2, with simple algebraic simplifications from (2) it is obtained:

$$m = \frac{2hf}{c^2}$$

For (1) we can replace $2hf$ with E, so we have:

$$m = \frac{E}{c^2}, i.e.: E = mc^2$$

Therefore, the energy E irradiated by body B is equal to the mass loss suffered by B after the emission, multiplied by the square of the speed of light in the vacuum.

This derivation is a didactic simplification that we need to understand how the relation $E=mc^2$ can be derived from experimental information that gives information about energy, that is, by the law $E=hf$ which expresses the energy of a quantum of irradiated light. Since energy as a magnitude is defined as acceleration impressed to a mass and multiplied by a space, the information given by $E=hf$ contains information on the mass concept, which can be made explicit in the deduction chain we have seen. We shall then see that this observation for which there is information concerning the concept of mass in the premises is relevant, but we shall also see that the simplified Rohrlich derivation hides an important problem.

4.5   E=mc² and increase of mass with speed

In the time when several physicists, including Einstein, came to conceive and hypothesize, if not to demonstrate, the relation $E=mc^2$, some experimental data were provided by the first particle experiments carried out with elementary processes and equipment, but conceptually identical to today's particle accelerators. It was observed that by increasing the speed of particles that were accelerated by electric currents, the energy needed to obtain additional speed increments was greater than the expected one, showing that as a result of the acceleration suffered the mass of the particles had increased.

The increase of mass with speed is therefore a phenomenon observed at the beginning of particle physics, then tested experimentally, and finally assumed as a fundamental principle in building accelerators. What is the quantitative relationship between mass and speed? This amount is expressed by a lesser known relationship, but as important as $E=mc^2$, which we shall see in the next section.

4.6   *Mass, energy and speed: non-relativistic derivation by Lewis (1908)*

To understand the relationship between mass, energy, and speed in a bit more specialist way, let us now see some parts of a 1908 paper titled *A Revision of the Fundamental Laws of matter and Energy*[38], whose reading is accessible under the usual condition of recalling school physics. The author is Gilbert Lewis, author of fundamental studies on chemical bonds (see Encyclopaedia). The interest of this paper is not only historical, and we do not read it at all for the pleasure of erudition. We shall read it, aware that it is a non-current exposure of the problem and corresponds to the way it was meant at another time, because in it we shall find an exposition that makes the relationship between mass, energy, and speed intelligible, but also shows the root of the problem that inevitably binds the theory of this relationship to special relativity.

In the original text the letter V is used for the speed of light, but here it will be replaced by the usual symbol c. There are some footnotes by the author, reproduced for simplicity after each paragraph. Only the parts of the paper strictly relevant to our purpose are reproduced.

> Recent publications of Einstein [2] and Comstock [3] on the relation of mass to energy have emboldened me to publish certain views which I have entertained on this subject and which a few years ago appeared purely speculative, but which have been so far corroborated by recent advances in experimental and theoretical physics that it seems desirable to subject these views to a strict logical development, although in so doing it will be necessary to modify those fundamental principles of the mechanics of ponderable matter which have remained unaltered since the time of Newton.
>
> [2] *Ann. Phys.* xviii. p. 639 (1905).
>
> [3] Phil. Mag xv. p. 1 (1908)
>
> The recent experiments which indicate a change in the mass of an electron with the speed, together with the phenomenon of radioactivity, have in some minds created a doubt as to the exact validity of some of the most general laws of nature. In the following pages I shall attempt to

---

[38] See Lewis 1908.

4. The relativistic concept of mass: E=mc²

show that we may construct a simple system of mechanics which is consistent with all known experimental facts, and which rests upon the assumption of the truth of the three great conservation laws, namely, the law of conservation of energy, the law of conservation of mass, and the law of conservation of momentum.

$(.\ .\ .\ .\ .\ .)$

The Relation of Mass to Energy.

When a black body [4] is placed in a beam of light it is subject to a pressure or force which tends to move it in the direction in which the light is moving. If $\dfrac{dE}{dt}$ denotes the time-rate at which the body receives energy, f the force, and c the velocity of light, we have in rational units the formula

$$f = \frac{1}{c}\frac{dE}{dt}.\,(1)$$

This important equation, which was obtained by Maxwell as a consequence of his electromagnetic theory, and by Boltzmann through the direct application of the laws of thermodynamics, has recently been verified with remarkable precision in the beautiful experiments of Nichols and Hull. [5]

[4] In place of a black body we might consider a partially reflecting one. The equations thus obtained are more complicated but lead also to the simple equation (7).

[5] Phys. Review, xvii. pp. 26 and 91 (1903).

A body subjected to the pressure of radiation will acquire momentum, and if we are to accept the law of conservation of momentum, we must conclude that some other system is losing in the same direction an equivalent momentum. We are thus led inevitably, as Poynting has shown, to the idea that the beam of radiation carries not only energy but momentum as well.

The body subject to the constant force of radiation f, will obviously acquire momentum at the rate

$$\frac{dM}{dt} = f.\,(2)$$

Combining equations (1) and (2) gives

$$\frac{dE}{dM} = c.\,(3)$$

The ratio of the acquired energy to the acquired momentum is equal to the velocity of light. The beam of radiation must, therefore, possess energy and momentum in the same ratio. Hence for the beam itself, or any part of it,

$$\frac{E}{M} = c. \ (4)$$

To anyone unfamiliar with the prevailing theories of light, knowing only that light moves with a certain velocity and that in a beam of light momentum and energy are being carried with this same velocity, the natural assumption would be that in such a beam *something possessing mass moves with the velocity of light and therefore has momentum and energy*. Notwithstanding its apparent divergence from the commonly accepted light theories, I propose to adopt this view and see whither it leads.

Postulating the validity of the fundamental conservation laws mentioned above, we shall need in the following development only this one cardinal assumption, that a beam of radiation possesses not only momentum and energy, but also mass, travelling with the velocity of light, and that a body absorbing radiation is acquiring this mass as it also acquires the momentum and the energy of the radiation. Therefore a body which absorbs radiant energy increases in mass.

The amount of this increase is readily found. If in general we write momentum as the product of mass and velocity, then the momentum of any part of a beam of radiation having the mass m will be given by the equation:

$$M = mc. \ (5)$$

The increase dM in the momentum of the body absorbing the radiation will therefore equal the increase dm in its mass, multiplied by the velocity of light,

$$dM = c \, dm. \ (6)$$

Combining this equation with (3) we find

$$dm = \frac{dE}{c^2}, \ (7)$$

or if we write $c = 3 \times 10^{10}$ centimetres per second,

$$dm = 1{,}111 \times 10^{-21} \, dE.$$

Thus a body receiving or emitting radiant energy gains or loses mass in proportion and by the amount 1.111×10^{-21} grams for every erg. This is a small quantity, indeed, but one which is not to be neglected.

Assuming the fundamental conservation law, we must regard mass as a real property of a body which depends upon its state, and not upon its history. Hence it is obvious that if in any other way than by radiation the body gains or loses energy it must still gain or lose mass in just the above proportion. In other words, any change in a body's content of energy is accompanied by a definite change in its mass, regardless of the nature of the process which the energy change accompanies.

(. . . . . .)

Since therefore when a body loses a given quantity of energy it always loses a definite quantity of mass, we might assume that if it should lose all its energy it would lose all its mass, or, in other words, that the mass of a body is a direct measure of its total energy, according to the equation,

$$m = \frac{E}{c^2}. (8)$$

We should then regard mass and energy as different names and different measures of the same quantity, and say that one gram equals 9×10^{20}ergs in the same sense that we say one metre equals 39.37... inches.

(. . . . . .)

Until here, Lewis deduced the relationship between mass and energy from an electromagnetic law, expressed in equation (1), which was confirmed by experiments made by Nichols and Hull, that had shown that light exerts mechanical pressure. The equation (8) obviously equates to:

$$E = mc^2. (9)$$

Then Lewis moves to a hypothesis about the relationship between mass and speed:

The axiom which we must surrender is the one which states that the mass of a body is independent of its velocity. We have concluded that mass is proportional to content of energy. When a body is set in motion it gains kinetic energy and therefore its mass must change with its velocity. In place of the axiom which we have abandoned we must substitute equation (7).

Before investigating the consequences of this step it is necessary to define exactly the principal mechanical quantities which we are to use.

Extension in space (l) and time (t) will be measured in the usual way and the centimetre and the second will be employed as units.

Force (f) will be given its usual significance and the unit, the dyne, will be that force which, acting upon the International standard kilogram, when the latter is at rest, imparts to it an initial acceleration of 0.001 *cm/sec²*.

The momentum (M) of a moving body will be measured by the time in which it is brought to rest under the influence of a constant opposing force of one dyne acting in the line of its motion.

The mass (*m*) of a moving body will be defined as the momentum divided by the velocity (*v*), that is,

$$m = \frac{M}{v}. (10)$$

The limiting ratio of the momentum of a body to its velocity, when it is brought to rest, will be called its mass when at rest. The unit of mass is the gram.

The kinetic energy (E') of a body will be measured by the distance through which it will move before being brought to rest by a constant opposing force of one dyne, acting in the line or the body's motion. The unit of energy will be the erg.

These definitions, although somewhat unusual in form, are perfectly consistent with the ordinary definitions of Newtonian mechanics. But they have been so chosen as to be consistent also with equation (7) and the fundamental conservation laws. Obviously equation (7) itself is not inconsistent with these conservation laws, for although a body increases in mass as it gains kinetic energy, some other system is losing the same mass as it loses the same energy.

In accordance with the above definitions we may write

$$dM = f\, dt, (11)$$
$$dE' = f\, dl. (12)$$

Let us consider a body originally moving with a velocity *v* subjected for the time *dt* to a force *f* in the line of its motion. Its momentum and kinetic energy will change according to (11) and (12) by the amounts

$$dM = f\, dt,$$
$$dE' = f\, dl = f\, v\, dt.$$

Hence

$$dE' = v\, dM. (13)$$

So far the equations are those of Newtonian mechanics, but now in substituting for M from equation (10) we must regard *m* as a variable and write

$$dE' = m\, v\, dv + v^2\, dm. (14)$$

4. The relativistic concept of mass: E=mc²

This will be our fundamental equation connecting the kinetic energy of a body with its mass and velocity.

Introducing now the relation of mass to energy given in equation (7) we may write,

$$dE' = c^2 dm,$$

and combining this equation with (14) gives

$$c^2\, dm = m\, v\, dv + v^2\, dm.$$

This equation, containing only two variables, m and v and the constant c, may readily be integrated as follows. By a simple transformation

$$\left(1 - \frac{v^2}{c^2}\right) dm = \frac{mvdv}{c^2}.$$

Writing $\beta = v/c$, and noting that

$$\frac{v\, dv}{c^2} = \frac{1}{2} d(1 - \beta^2),$$

we see that

$$\frac{dm}{m} = -\frac{1}{2}\frac{d(1 - \beta^2)}{(1 - \beta^2)}.$$

Hence

$$\log m = -\frac{1}{2}\log(1 - \beta^2) + \log m_0$$

where $\log m_0$ is the integration constant. Therefore

$$\log\frac{m}{m_0} = \log\,(1 - \beta^2)^{-\frac{1}{2}}$$

or

$$\frac{m}{m_0} = \frac{1}{(1 - \beta^2)^{\frac{1}{2}}} . (15)$$

This is the general expression for the mass of a moving body in terms of β, the ratio of its velocity to the velocity of light. When β is zero, $m = m_0$. m_0 represents therefore the mass of the body at rest.

In this way, Lewis's development approaches to the relationship that was sought: in it we find Lorentz's factor, though applied here in relation to mass, not space and time, and derived from a completely different path from that of relativity. In fact, (15) is exactly equivalent to:

4. The relativistic concept of mass: $E=mc^2$

$$m = m_0 \, \frac{1}{\sqrt{1 - \dfrac{v^2}{c^2}}}$$

or

$$m = m_0 \, \gamma$$

and this is the relationship always applied to the mass increase of particles in accelerators since the time of Lewis's paper.

Lewis continues with a discussion of the result and presenting of some experimental results very consistent with the hypothesis:

> If we substitute in the equation numerical values of β we find that, while the quotient m/m_0 becomes infinite when the velocity equals the velocity of light, it remains almost equal to unity until the velocity of light is closely approached. Thus a ton weight given the velocity of the fastest cannon-ball would, according to this equation, gain in mass by less than one millionth of a gram. It is obvious that, except in those unusual cases in which we deal with velocities comparable with that of light, our non-Newtonian equations are identical with those of ordinary mechanics far within the limits of error of the most delicate experiments.

$(\ldots\ldots)$

> In a series of remarkably skilful experiments Kaufmann [8] was able to measure the ratio of electric charge to mass (e/m) for negative particles moving at different speeds. Assuming that the charge is constant, the fact that e/m varies with the speed of the particle must be attributed to a variation of the mass with the speed. On this assumption it is possible to calculate from Kaufmann's experiments the values of m/m_0 at the different velocities.

> [8] *Phys. Zeit.* iv. p. 54 (1902); *Ann Phys.* xix. p. 487 (1906).

> The mass of a negative particle is usually spoken of as electromagnetic mass, but if we are to hold to our definitions we must recognize only one kind of mass. In general we have defined the mass of a moving body as the quotient of the time during which it will be brought to rest by unit force, divided by the initial velocity. It matters not what the supposed origin of this mass may be. Equation (15) should therefore be directly applicable to the experiments of Kaufmann. In the following table are given the values of

4. The relativistic concept of mass: $E=mc^2$

m/m_0	β (observed)	β (calculated)
1	0	0
1.34	.73	.67
1.37	.75	.69
1.42	.78	.71
1.47	.80	.73
1.54	.83	.76
1.65	.86	.80
1.73	.88	.82
2.05	.93	.88
2.14	.95	.89
2.42	.96	.91

m/m_0 found for the different observed values of β in the second column. The third column shows those values of β which would correspond with the same values of m/m_0 according to equation (15).

It will be seen that the observed values of β follow to a remarkable degree the same trend as those which are calculated by equation (15), but are in every case six to eight per cent higher.[9] I believe that these differences lie within the limits of experimental error of Kaufmann's measurements. It is true that he claims a higher degree of accuracy, but, notwithstanding the extreme care and delicacy with which the observations were made, it seems almost incredible that measurements of this character, which consisted in the determination of the minute displacement of a somewhat hazy spot on a photographic plate, could have been determined with the precision claimed. Moreover, Planck [10] and Stark [11] have pointed out certain corrections which probably should have been made by Kaufmann and which would produce a material change in his results.[12]

[9] The constancy of the difference between the observed and calculated values of β is striking, and would alone indicate some constant error in Kaufmann's results.

[10] *Verhandlung Deutsch. Phys. Ges.* ix. p. 301 (1907).

[11] *Ibid.* x. p. 14 (1908).

[12] In reply to Planck see Kaufmann, *ibid.* ix. p. 667 (1907).

In the latter part of Lewis's paper there are the considerations we know about Lorentz's gamma factor behaviour: applied to mass increase, it indicates that any mass would become huge near the speed of light, which cannot be achieved because the mass would become infinite. This result is experimentally verified in particle accelerators: it is today, but it was already, roughly, in primitive particle experiments that could be performed in the early twentieth century. Under the conditions of mechanical speeds of ordinary experience, where every speed is very small with respect to light, the mass increase with speed is so small that it cannot be observed and it has no practical or technical relevance.

4.7   Mass, energy and speed: problems revealed by non-relativistic conception

We have seen that without resorting to relativity and without invoking the special assumption that light speed is constant for each observer, two formulas can be obtained, one of which gives us the equivalence between energy and mass, and the other the increase of mass with speed. The two formulas could also be derived with a relativistic technique from Einstein's postulates, but we shall deal with this later. With the derivations by Rohrlich and Lewis we have neither a modern nor a complete theory of the relationship between energy and mass: to achieve this result we should deepen the study of particle and nuclear physics. The two derivations, a very simplified and a primitive one, however, serve to understand how the relationship between energy and mass can be obtained rationally and to precisely define the relation of these formulas with the theory of relativity. What we must now observe is that in both formulas:

$$E = mc^2$$

and

$$m = m_0 \, \frac{1}{\sqrt{1 - \dfrac{v^2}{c^2}}}$$

or

$$m = m_0 \, \gamma$$

the speed of light appears as a constant. But constant with respect to what? If we think of the physical meaning of these formulas, it is clear that it is necessary to either go back to the conception of the ether and consider c constant with respect to the ether, or adopt Einstein's postulate and consider the constant c as absolute in each inertial system.

In the first case, the two formulas would be approximated for the reason that we do not know the speed of the Earth and any other system with respect to the ether. In particular, it would be incorrect to calculate the mass increase of an accelerated particle by applying speed v of the particle inside the accelerator, since instead we should determine the particle speed with respect to the ether, summing or subtracting the speed of the Earth and the accelerator with respect to the ether.

In the second case, by considering the speed of light c as absolute constant, the two formulas assume an exact meaning. The same is true in the context of Lorentz's hypothesis, where c is also constant for every observer: let us never forget this.

What we cannot do if we want to preserve the two formulas is to adopt the simple ballistic theory and assume that the speed of light is constant only with respect to the emitting body, and that it can then be added to the speed of the emitting body, because this hypothesis would lead to paradoxical consequences.

In fact, the formula $E=mc^2$ expresses the energy released by a reaction in which the mass is lost. If speed c is not an absolute constant, this amount would not be invariant: we would imagine that the energy released by a nuclear reaction is different if the reaction takes place in a system at rest or in a moving one, which, by analogy with a non-nuclear reaction, is like imagining that the combustion of a match releases different amounts of heat if it occurs either on the Earth at rest, or inside a ship, or in a flying aircraft, and so on. But since the law of inertia tells us that the state of rest cannot be distinguished from that of rectilinear motion at constant speed, we cannot accept this conclusion. And in any case, if the speed of light were no absolute constant, but constant only with respect to the emitting body, we could not write a relationship where it is considered and used as a physical constant. The expression $E=mc^2$ would not have any definite sense in this hypothesis.

As for the expression $m=m_0\gamma$, things are even more paradoxical. We know that if v tends to reach the speed of light, then the gamma factor tends to infinite, if it reaches the speed of light then it becomes senseless because the result of the division of a number by zero is undefined, and if v exceeds c then the gamma factor becomes a complex number, root square of a negative number, whose physical meaning we would not know how to interpret.

In this case, let us imagine an accelerated particle at a speed v just a little lower than the speed of the light. The mass of this particle with respect to the accelerator is increased by hundreds of times. The speed v we are talking about is measured with respect to the accelerator, of course, but there is no reason to believe that it is a privileged speed, that is, the absolute speed to take as main reference. So now let us imagine that there is a man walking towards the particle outside the accelerator and that the speed v is added to that of man, and so it approaches that of the light so close that the gamma factor, considering the speed of the particle with respect to the man, becomes something like a hundred million billion. Since in the formula $m=m_0\gamma$ the speed c is constant, applying the particle speed with respect to the man rather than that to the accelerator in the formula, we would calculate that the particle reaches a mass of a number of kilograms with tens of zeros, and hence perhaps it would break the Earth's crust destroying the planet, or who knows what else would happen. If the speed of the particle exceeded that of light, it is unthinkable what should happen. But since each speed different from that of the light is relative to the frame chosen to calculate it, there is no reason to assume as good for calculating the speed of the particle the speed in the accelerator rather than that in any other frame, and hence if we do not assume the speed of light as absolute constant, we cannot use the formula $m=m_0\gamma$, which however seems to be consistent with the experimental data. Today it is believed that a speed of 0.999999999976 c or even higher for the Higgs boson has been reached. These speeds are lower than c just one centimetre per second, so in the hypothesis of emission theory, the formulas $E=mc^2$ and $m=m_0\gamma$ could not have the form they have, and the theory leading to their deduction should be reconsidered completely, not unlike the theory that leads to Maxwell's laws. Both formulas, on the other hand, can conserve their form both in Lorentz's hypothesis and in Einstein's relativity; in Lorentz's case with no logical problems of any kind, in Einstein's case provided that we can find the way to interpret and use the theory in a not conventional and not apparent way.

5. *Time and space dilatation and law of inertia*

5.1 *Rewording the problem*

Before proceeding, it is time to express a statement that the readers will have to take into account, concerning the readers and their relationship with relativity. We have seen that relativity is subdivided into relativistic kinematics that relates to spatial and temporal dilatation in relation to speed, and into relativistic dynamics concerning the two formulas that radically innovate the mass concept. As for the second part, the dynamic one, those who are not specialists must confine themselves to obtaining a general idea of the relationship between mass and energy. Following a purely archaic paper like that of Lewis, the relationship ceases to belong to the world of mystery and we understand how it can be derived rationally, but this remains necessarily within the limits of an elementary comprehension of the problem: in order to become experts it would be necessary to know particle and nuclear physics, and in the absence of such knowledge readers are not allowed to judge, but only to detect open issues that can be understood by anyone, such as the problem of the reference frame for mass gain. As for the kinematic part, the situation is very different. Probably the readers are already aware of it, but it is good to observe explicitly that if readers have sufficient knowledge to understand Lorentz's hypothesis, to follow the few algebraic steps with which formulas are derived, to understand that the constancy of c for all observers is the logical result of Lorentz's hypothesis, to realize the difficult and peculiar consequences of Einstein's postulate about the constancy of c even in the absence of ether, and finally to understand the problem of the invariance of Maxwell's equations considering them as any function of x,y,z,t (we saw that it is not necessary to know in detail the meaning of Maxwell's equations to understand the invariance problem), then people who are able to understand these things have transformed themselves into experts in relativistic kinematics, and possess all the prerequisites for judging the kinematic part of Einstein's theory with their own intelligence. There is no more relevant or specialized knowledge than that outlined here. There are no behaviours of nature known by specialists and unknown to the general public which would make Einstein's specific postulate more plausible tind would solve the dilemma of the conventional, apparent, or realist interpretation of the theory in Einstein's version. There are no experimental facts or superior mathematical constructions that can make sense of how the realist interpretation of temporal dilatation, which Einstein makes dependent on the infinite relative speeds that everything in the world has with respect to infinite other things, could yield logically consistent results. So with regard to the kinematic part — and obviously only the kinematic part — the readers who understand what has been described up to here know the problem as experts, no less than Feynman and other Nobel Laureates, and have the right to judge the problem with their own intelligence considering themselves skilled at the highest level.

This said, until here we have seen that the question of the relationship between mass and speed has given us an excellent reason to wish to find some logical ground that allows us to accept the postulate of the absolute constancy of c. But the detailed analysis of the consequences of the postulate that we have developed above has shown us that the conventional or apparent interpretation of the postulate would be useless, because it would give a fictitious solution to the problem of the invariance of Maxwell's laws and any other problem related with the use of c as a physical constant, including the problems of relationship between energy and mass and speed. The realist interpretation, on the other hand, is incompatible with the law of inertia, because given the law of inertia it is impossible to decide, given two systems in relative motion, in which time would dilate, and in addition given the law of inertia each system would have infinite dilatations with respect to infinite systems in motion with respect to it at different speeds, that exist at each instant.

By discussing the problem, we have rigorously defined the meaning of the conventional, apparent and realist interpretations. But since the words "conventional", "apparent" and "real" in the ordinary sense are equivocal and can be discussed infinitely, we shall now introduce a simplification that consists of bringing together the conventional and apparent interpretation into one family, and in defining a schematic and unique criterion to distinguish this family of interpretations from the realist one.

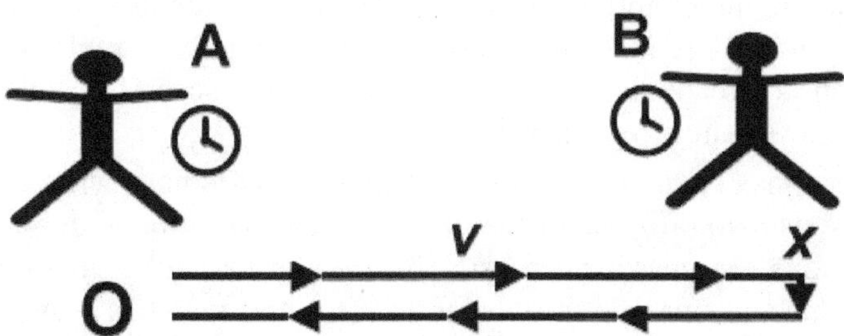

There are two people, A and B, who have two identical clocks. When the two people are in O, they set their clocks to mark the same time without the need for remote synchronization procedures. Or (if we like it better) they adjust their clocks by means of signals, and synchronization gives the same result, because the two people are at relative rest. Then B travels at constant speed v up to point x, there reverses the sense of motion and travels back to O. A and B do not activate any synchronization of their clocks while B is in motion.

Given everything we know, we can formulate the following statements that are not said to be true or plausible, but in which given Einstein's premises the words have recognizable meaning:

(0) while B is in motion, if A wanted to determine the time marked by clock B using electromagnetic signals, it would obtain a different measurement than indicated by clock B.

(1) when A and B come back together, or are again in relative rest, the two clocks agree because the proper time of system B has not changed. Only time possibly measured by A remotely would have a different numeric value.

(2) when A and B come back together, the two clocks do not agree because the time of system B has dilated as a result of the motion condition. B's time durations were contracted, and B's clock is late compared to that of A.

The same is true for space measurements, and there is no need to specify cases also for space. Suppose we accept assertion (0). Since the premises are those of relativity, all agree that consequently discordant time and space measures would be correlated by Lorentz's formulas, and that these are the key to correct the measure once assertion (0) is accepted. Until here, the theory expresses nothing but the prescription of taking Lorentz's factor into account when carrying out remote measurements by means of signals: this prescription is already as problematic as Einstein's second postulate is, but we now assume to accept it.

The assertions (1) and (2), on the other hand, contradict each other, and require decision for one or the other. We cannot agree with both. Two decisions are possible, and for this we define as the *apparent interpretation* the one which decides for (1), and we define as the *realist interpretation* the one which decides for (2).

The apparent interpretation, which at this point absorbs the conventional one, is useless, and gives rise to fictitious solutions. What changes in natural processes by the fact that people in A at rest, if they wanted, would measure the time of B in motion considering it dilated, if they did not consider Lorentz? We have already discussed this issue in detail.

The realist interpretation (which is conceivable under the hypothesis of the absolute frame of ether) is impossible, because given the law of inertia nothing can prevent us from considering B at rest and A in motion, or A at rest and B in motion, so we can only express ourselves in open contradiction by saying: when A and B join together, A's clock is late with respect to the clock of B, and B's clock is late with respect to that of A. Moreover, if there were a third system C moving with respect to each at a different speed, then we would have different contractions and dilatations for AB, AC and BC relationships, and many more if there were a fourth system D, and so on, endlessly.

We shall see that the first problem, that of symmetry, is treated by the literature, and has been subject to difficulties and controversy. The second problem, that of infinite multiplicity of contractions, is not usually detected, probably due to the exhaustion of energies determined by the discussion of the first one. I did not find any trace in the literature, except in one case of the very first time which we shall meet in the following: yet the attentive reader should remark at a glance that although the examples in the books always take into account only two systems to construct the schematic representation of the phenomenon, it is obvious that this way of expressing things is always and only a simplification for expression purpose, because the physical world is made up of countless systems in relative motion.

5. Time and space dilatation and law of inertia

The problem is dramatic. In fact, as long as the known problem was only that of the invariance of Maxwell's laws, one could say: why should we take this hypothesis, which assumes c as absolute constant and comes into conflict with the law of inertia? By discarding the hypothesis of the ether, the first thing to do would be to explore the hypothesis of emission, i.e. to assume that c is constant with respect to the source of emission of electromagnetic waves, and to review the whole process by which the laws of electromagnetism were obtained, and to reformulate them consistently with the hypothesis that light inherits the inertia of the emitting body, as is hypothesized for anything else. If this succeeded, Maxwell's equations would take another form, but c, the typical speed of light, being understood as the speed value with respect to the emitting body, would no longer be considered a physical constant.

But just as these problems were at the centre of attention, in the very same era in which the hypothesis of special relativity was formulated, two other fundamental laws were formed, $E=mc^2$ and $m=m_0\gamma$, in which c is constant and can be interpreted in a sensible way only as an absolute constant. So if there were not the theory of special relativity (or Lorentz's), an immeasurable amount of work should be done to reinterpret the phenomena and experimental data that led to Maxwell's laws and particle physics, perhaps questioning the notions of mass, space, and time more radically than anyone has ever done, including Einstein. Or we should go back to hypothesizing the ether (and Lorentz's hypothesis), whose existence no experiment has ever revealed.

But the theory of special relativity is no solution because it determines a condition of contradiction of all measurements of space and time. Competitive theories have difficulties and are incomplete, but Einstein's relativity is in a worse state because it forces us to choose between an insignificant interpretation and an openly contradictory one.

We can summarize the problem in the following table, in which we consider the ether theories without and with the Lorentz hypothesis, the emission theory and special relativity, and we point out the state of problems relating to space and time measurements, Maxwell's laws, and the relationships between energy, mass and speed. The table shows the status of the three problems in the different scenarios:

Theories	Problems		
	Space and time measures	*Maxwell*	$E=mc^2$ $m=m_0\gamma$
Ether	OK	Approximation. Disagreement with observation because the speed in the ether is not observed.	Approximation. Disagreement with observation because the speed in the ether is not observed.
Ether with Lorentz's correction	OK	OK	OK
Emission	OK	OK if fundamental equations were rewritten, else approximation.	Irremediable approximation for the increase of mass in particle accelerators. Formulas should be modified radically.
Special relativity	***Irremediable contradiction***	OK if the theory were consistent	OK if the theory were consistent

Looking at things from a strictly logical point of view, and without professing any opinion about physical problems, the only theory consistent with Maxwell's equations and with the relationship between mass and energy is Lorentz's old theory. The ether theory without Lorentz's correction as well as the emission theory would require the re-formulation of the other theories, but they were not made impossible by any contradiction, while Einstein's theory is flawed at the root of a contradiction that makes it useless.

5.2   Nature of contradiction

What exactly is a contradiction? The question seems trivial, but we now need a brief explanation of the logical and the psychological character of the contradiction.

From the logical point of view, often the mistake is made of relating the concept of contradiction to temporal considerations, whereas contradiction concerns only elementary assignments of anything to one or another set. Contradiction consists of incompatible, mutually exclusive propositions correlated with the conjunction *"and"*, which means that the assertions are both true. When we say "A is B *and* A is not B", it is not necessary to add "at the same time" so that there is contradiction. Generally, the reference to time is added because it is known that anything A can be B at a given time and can be not B at another time, but in this case the speech becomes more articulated; contradiction is simpler than this. Just consider some elementary mathematical examples to realize that time is not in question: the expression "this triangle is equilateral *and* this triangle is right" is contradictory because by Pythagoras we know that right triangles cannot be equilateral and this has no relation to the dimension of time. The expression "2+2=4 *and* 2+2=5" is contradictory because the problem "2+2" has only one solution, unless some new mathematics is conceived; even here time is not in question.

Contradiction is a judgment that violates elementary rules of attributing any entity to two sets. If we represent things graphically as well as in words we see that we have four ways in which two sets can be related:

- they can be entirely in one another, as when we say "all men are mammals":

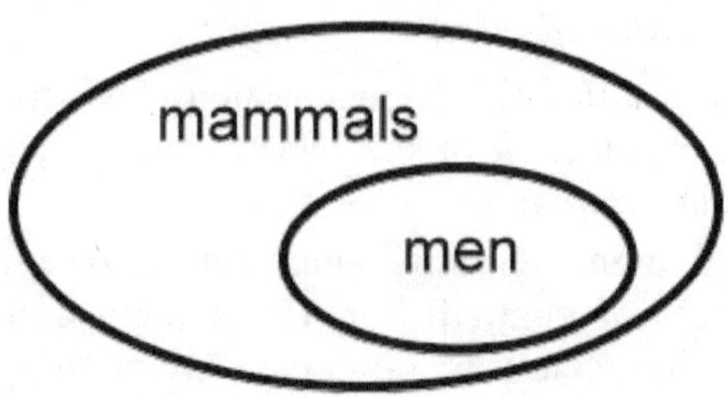

- they can be entirely coincident, as when we say "all platypuses have a beak and no other mammal has a beak". In this case the same set is described in two different ways:

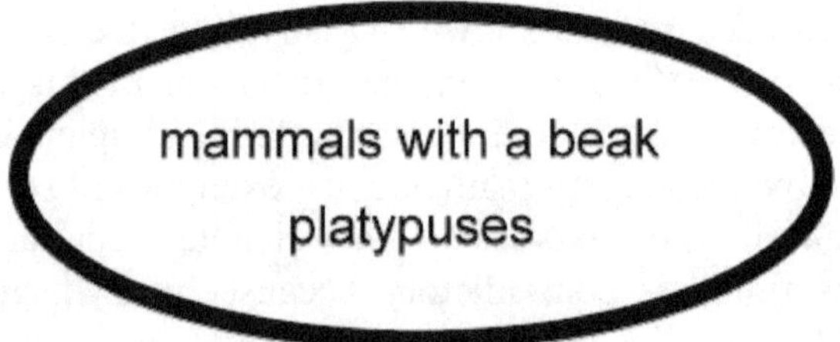

- they can have an intersection, as when saying "some mammals are capable of flying (bats) and some flying animals are not mammals (birds, insects)":

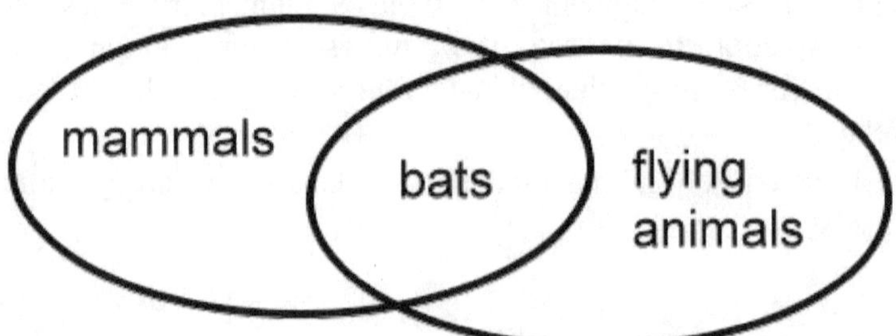

- they can be completely extern to one another, as when we say "no bird can write":

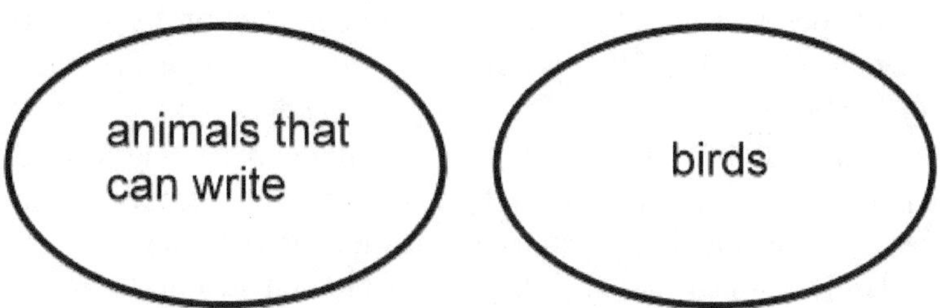

Contradiction consists of the violation of these elementary rules, and has the characteristic that it cannot even be expressed when the case with which to deal with is trivial. Suppose we have a canary in front of us and want to say "no bird can write *and* this canary can write." In words a contradiction can be pronounced or put on paper, though it is immediately recognised to be nonsense, but graphically it cannot even be expressed. In fact, let us try to draw the canary representing it with a symbol and to express the contradiction by placing this symbol in both figures representing the two sets: since there is no intersection between the set of the animals that can write and the set of birds, the canary can only be put in one of two, and not in both sets.

This example also shows us that contradiction can only exist in expression, and not in things: if we decided to physically put animals that know how to write and birds into separate cages, we could not put our canary in two different cages. We could put it in the cage of the animals that know how to write, but with that we would heal the contradiction, because not putting the canary in the birds' cage we would express the overall judgment: "no bird can write *and* this canary can write *and* this canary is not a bird" which is false in relation to experience, but has resolved the contradiction by adding the erroneous specification that consists of removing our canary from the set of the birds.

Contradiction can only exist within an articulated discourse, whose various parts are preserved by memory, and where the memory of the parts may be attenuated. If I contradict myself, it is because I first developed an argument that led me to a certain conclusion A, then I attributed to the complex of that argument a label for the use of memory, the word or periphrasis "A", then other considerations have led to further conclusions B that exclude A, then I called "B" the complex of the second argument, and finally I pronounced the words "A" and "B" and expressed the judgment "A *and* B" when I had an attenuated memory of the original context in which I came to conclusion A which is mutually exclusive with B. I express the contradiction because my intellect is not able to pay thorough attention to the complex of considerations that led me to conclusion A and also to the considerations that led me to conclusion B, incompatible with A. Contradiction does not exist in the world, but only in the human mind, which carries out its thoughts over time, and to carry out its thoughts needs memory, which however has limited capacities, and attenuates what is far away. Certain persons never contradict themselves because they are able to only have consciousness of what they are obsessively concentrated on in the present, and never express relationships between the present and past states of consciousness.

This recalls to the mind ancient philosophical problems: Duns Scotus and St. Thomas, who were subtle people, had observed that contradiction exists only in expression, and consequently the question had arisen if God could contradict himself, and they concluded that no, that he cannot, that the principle of non-contradiction is stronger than the same divine omnipotence. If you want, this is a questionable opinion in turn, and other theological schools have attributed to divine omnipotence also the power to act contradictions if so decided, but this no issue of interest for us now, because we are dealing with the logic that is possible for mankind. For men is valid the principle that a contradictory world is not conceivable, because contradiction exists only in human expression, which is at the mercy of the fragility of memory. Therefore, a contradictory world cannot enter into human experience, because even if it existed by divine decree, we men could not recognize it. When we face contradictory situations, we know that error is in our interpretation, not in things. Or at least, we know that if contradiction were in things we could not recognize it, but we would always continue to attribute it to our expression. We are talking here of course of strictly logical contradiction, that is, of violating the basic formal rules of reasoning, not of conflicts between things that have physical existence. The world has infinite conflicts, often improperly called contradictions, but it does not have formal contradictions.

Now the readers should do a psychology experiment on themselves, which usually gives a positive result. Focus on what you are reading, and read the phrase: "no bird can write, so this canary cannot write." Then read the sentence: "no bird can write *and* this canary can write." In addition to being clear that the first sentence is consistent while the second is contradictory, do you realize that when you pay attention to the first sentence, your state of consciousness has something like a different colour than when you focus on the second? That in the first case everything flows smoothly, while in the second case one feels a sort of emotional discomfort, but it disappears immediately seeing that everything is irrelevant because the judgment that declares the second phrase to be contradictory is absolutely certain? We notice a contradiction because a certain state of mind makes us conscious of it, but it is not said that this always happens. People unable to focus attention beyond the horizon of what interests them in the present instant, may never be able to evaluate the coherence of expressions of others and pronounce the judgments "this speech is consistent", "this is contradictory", because the state of mind revealing the consistency or contradiction of an expression is absent in them, and for them the problem has no meaning.

In addition, we men contradict ourselves for reasons of two very different kinds. Often we express contradiction because the complexity of the branches of a problem exceeds the ability of our mind to pay attention to everything involved in the study of that problem. This is the case of error in a strictly theoretical sense. But equally often we express contradiction because some parts of a problem are related to conflicts in our desires, and consequently coherence gives rise to disappointment, while contradiction receives a reward that the imaginative satisfaction of our desires assigns to us. The state of consciousness in which we fall in contradiction because our soul receives a prize from it has been studied by Leon Festinger[39], who called this phenomenon "cognitive dissonance", and described analytically the strategies by which we reduce the discomfort that dissonance implies.

---

[39] Festinger 1957.

5.3 The conflict between apparent and realist interpretation of relativity

At this point, we must boldly acknowledge that the relativistic literature since more than a hundred years is in a state of cognitive dissonance. The theory is contradictory, but contradiction is asserted because the theory is promising, and creates the expectation of different kinds of reward for those who accept the contradiction. Contradiction creates a state of uneasiness, but this is removed through self-deception strategies.

These words will appear overwhelming and incredible to those who have not yet experienced the study of special relativity, given the immense prestige the world has attributed to this theory. But those who are reading this book because in the past they tried to study the theory and confessed to themselves that they did not understand it, if they tried to find the coherence of any treatise on relativity before abandoning the effort, they will now recognize in our words the exact description of their experience.

Since a hundred years ago until today, the relativistic literature is reproducing the following assertions related with the conjunction *"and"*, which supposes all them true:

> The dilatation of space and time between two moving systems is a phenomenon that concerns the measures taken by one system on the other

and

> "in its own reference system" everything has unique and constant measures, called "proper time" and "proper space"

and

> the "proper" measures do not change as a result of the state of motion

and

> the contraction and dilatation of spaces and times do not concern the "proper" measures of any system in motion or relative rest

and

> given the contraction/dilatation, the measures of moving systems change due to the state of motion

and

the "proper" measures can change as a result of the state of motion

and

measurement changes are observable only by moving remote observers via electromagnetic waves

and

measurement changes can also be observed with direct local measures

and

no clock changes its frequency because of being in motion

and

clocks change their frequency because of being in motion

and

given the law of inertia no system can be privileged, so the clocks of two systems disagree reciprocally and apparently only in the measures taken remotely and in motion

and

this system A in relative motion has undergone time dilatation with respect to this system B in relative rest state, so clock A is later compared to clock B even in local measures

and

this system B in relative motion has undergone time dilatation with respect to this system A in relative rest state, so clock B is later compared to clock A even in local measures

and

so on endlessly ...

This is the state of the affairs. The expression "measured in its own reference system", synonymous with the word "apparent", which authors would like to avoid saying, has the lion's share of responsibility for creating ambiguity. The expression "real" is never pronounced; then, however, we are informed that clocks in motion are apparently delayed *and* delayed truly too, *and* that if compared with clocks at rest, they show different indications.

Simplified in such a form, the typical expressions of the relativistic literature are obviously contradictory. But since the generalization of the law of inertia makes the difference between local times depending on the state of motion or relative rest inconceivable, relativity theory is asserted through the expression of contradictory phrases distanced over time and arranged so as not to be immediately recognizable because of the complication of the expressive context.

People who have tried to study relativity and after a certain effort have abandoned the attempt will recognize in this description exactly what they have experienced. The discomfort that we experience is completely different from what we experience in other cases when we realize we are not understanding something. When we abandon a study because we think it is too difficult for ourselves or because we decide not to invest the necessary effort, first underestimated, we simply have the impression that the matter is too complex, and we attribute to ourselves the inability or lack of will to overtake the difficulties. In the common cases, a theory we fail to understand seems significant in the general conclusions and assertions we know about, but cannot be captured in detail because of our subjective limits. In the case of relativity, instead, one finds that what has been read does not require any upper mathematical competence and any physics knowledge that is not elementary, and what was to be understood consists only of examples that tell us about clocks in motion and trains that run and thunderbolts that fall and are seen in different conditions, and yet we experience the abandonment of the study because we are unable to grasp the key of the coherence of the whole matter. Discomfort, however, is different than in any other case: those who abandon the study of special relativity feel that there is something strange and that the accounts do not balance; but the immense prestige of special relativity leads us to fall back on making us small and humble and to end by attributing inadequacy to ourselves. Some people otherwise deem to understand special relativity perfectly and, consequently, deem to be able to explain it and to write new treatises about it. The experience of these people is completely different: it involves accepting the state of contradiction in exchange for the emotional reward that one receives by adopting the modes of expression and social behaviours appropriate to the case, and all falls into Festinger's theory of cognitive dissonance. To understand how this disconcerting situation has been determined, to whose existence we would not believe if we had not experienced the discomfort that we feel when we try to study the theory of relativity without success, we shall now see some moments in the history of the issue, and then we shall discuss a picture of the motivations that produced this immense misunderstanding, which is perhaps the most abnormal anthropological phenomenon of the twentieth century.

5.4   Interpretation of special relativity shortly after 1905

Einstein's three papers and the note on $E=mc^2$ were published in 1905 by a prestigious journal edited by Max Planck, *Annalen der Physik*, and the publication granted Einstein a credit opening and some teaching assignments in Switzerland and Prague. Einstein's prestige grew so rapidly that in the years around the Great War he could work on general relativity by occupying a very favourable position in Berlin, almost without office duties. The detailed narrative of the story can be read in Isaacson[40], and it provides useful insights into the success of the theory even to those who have no curiosity for Einstein's person. It is well known that Einstein's popularity exploded in 1919, when the results of some astronomical observations made in exceptional circumstances were published and were interpreted as confirming the theory of gravity contained in general relativity; these observations are not exactly a proof of special relativity, but the world wanted to interpret them as a demonstration of everything that came from Einstein. Before then, however, it was not true that Einstein was a dark figure, as sometimes is said: reading biographies we learn that his 1905 works were known to all the specialists who were able to understand the related problem.

The underlying problem was known to a few physicists; one can easily imagine that most people were little concerned with the existence of the ether and the invariance of the electrodynamics laws, but it must be added that most experimental physicists did not care either. There was, however, an elite, which included top-level specialists, who attributed great importance to the theoretical problems associated with the existence of the ether, and found it crucial to find the key to maintaining Maxwell's equations in their form[41].

Let us see what the perception of both the problem and Einstein's papers by specialists who knew the details of it was. First of all, we must note that Poincaré had spoken of all the themes at stake in a

---

[40] Isaacson 2007 Chapter 9.

[41] Details about this can be read in Dingle 1972.

lecture[42] titled "L'état actuel et l'avenir de la physique mathématique" ("The present state and future of mathematical physics"), held in St. Louis, Missouri, in September 1904 and which had undergone many reproductions in different journals and books. Among other things, it is likely that Einstein knew this text already in 1904[43]. So people up to date, who knew the current state of science with respect to these problems, read ED1905 with these assumptions:

- In Einstein's first paper they had read a plausible hypothesis as an explanation of the phenomenon of Brownian motion. Poincaré had pointed to this issue as crucial in the 1904 lecture.

- In the second paper they found that the theory of photoelectric and electromagnetic quantum effects was a plausible conclusion of Max Planck's research.

- In the appendix to ED1905 they found the formula $E=mc^2$, but the specialists at the time knew that the hypothesis of transformability between mass and energy was plausible given recent experiments suggesting mass gain, and as we saw, Lewis for example developed the subject in 1908 also citing Einstein among his sources. The possible equivalence between mass and energy had been mentioned in generic terms by Poincaré in the 1904 lecture.

- The specialists knew the perennial problem of the ether.

- They knew the problem of Maxwell's laws invariance.

- They knew that Lorentz's hypothesis was artificial and considered such by its own creators, but they had been prepared to accept the idea that time may be local and not universal: this idea, hypothetical but devoid of illogical consequences in Lorentz's and Larmor's formulation, before them had never been conceived, while now it was known to those who knew the problem.

- They knew that Poincaré took seriously Lorentz's hypothesis, and had spoken about it extensively, also at the 1904 lecture. Among other things, in that lecture Poincaré expressed himself with the

---

[42] See Poincaré 1904.

[43] Miller 2001, p. 185.

rhetorical question[44]: "Should we not also endeavor to obtain a more satisfactory theory of the electrodynamics of bodies in motion?", and in his work he had explicitly mentioned also the clock synchronization technique by means of reflected electromagnetic signals and had dealt with its consequences in relation to Lorentz's hypothesis.

- But most of all, to the contemporary readers, the assertion that the speed of light should be constant for every observer did not seem so strange as it appears to those who come across it the first time. In fact, we know that the assertion is incomprehensible in the emission theory, or ballistic, for beams emitted by a source in motion with respect to the measuring apparatus; but on the contrary Lorentz's hypothesis about the theory of ether predicts that light speed measurement is constant for every observer, and makes it perfectly conceivable.

With these premises, specialists who were reading the papers of 1905 acknowledged the words of a promising young man who had mastered the electromagnetic argument and talked about it with professional competence, and that on a few pages that were scarcely understandable related his speculation to the Lorentz's hypothesis and suggested that this could be transformed into a much more elegant and not artificial theory. Instead of hypothesizing physical contraction while moving in the ether, Einstein suggested that the problem stems from a consolidated misconception of space and time measurement. Lorentz suggested with all clearness that things might be subject to a contraction of lengths and durations not proven by experience, while Einstein suggested confusedly but with a strong impact that the problem arose from a conceptual mistake consolidated for centuries, and that therefore the contraction could be defined and demonstrable with mere logical and mathematical considerations, with no need of any experience to confirm it. If Einstein's point of view had been consistent, a theorem would replace a hypothesis about nature and certainty would take the place of conjecture. This is the reason why specialists at that time not only were eager to listen carefully to Einstein's proposal, but had a

---

[44] Poincaré 1904.

predisposition to accept it: Einstein's advantage with respect to Lorentz was immense, because Einstein offered mathematical certainty to replace a contingent conjecture that was hard or impossible to demonstrate through experiments.

Chapters 2 and 3 of ED1905, which we have considered word by word, are difficult to read not only for those who do not know the problem, but for anyone, because they only contain a sketch of a theory expressed in a summary form. There is no reason to imagine that the greatest physicists of the time grasped the meaning of the theory in a more precise and accurate way than any other reader, because their competence prepared them to accept the theory for the advantage that it seemed to have on Lorentz's, but this was not enough to make Einstein's expression clear. Simply, those who knew the underlying problem had the impression that there was something demonstrable that could replace the unproven and artificial (but not illogical) hypothesis of Lorentz with a point of view devoid of physical hypotheses, which consisted only of eliminating prejudices and correcting methodological errors that had been consolidated by use and habit.

This was the perception of the theory, or rather of the potentiality of the theory. Given this premise, the theory of special relativity quickly gained popularity in the generation of younger theoretical physicists. Obviously we are talking about the very few who could frame the problem, but these very few were an elite, the founders of the imminent great reformation of the methods of physics in the twentieth century. In general, the elders resisted mentally, and Michelson, Lorentz, and Poincaré never accepted special relativity[45], yet they never managed to frame the misunderstanding with simplicity and describe the contradictory nature of the new theory. Poincaré might have succeeded, but he suddenly died in 1912. As for Lorentz, in 1909 he published the lessons on the electron he had held in previous years, and in this volume we read these perplexed considerations[46]:

I cannot speak here of the many highly interesting applications which

---

[45] Canales 2015 gives detailed information.

[46] Lorentz 1909 and 1916, p. 229-230.

5. Time and space dilatation and law of inertia

> Einstein has made of this principle. His results concerning electromagnetic and optical phenomena (...) agree in the main with those which we have obtained in the preceding pages, the chief difference being that Einstein simply postulates what we have deduced, with some difficulty and not altogether satisfactorily, from the fundamental equations of the electromagnetic field. By doing so, he may certainly take credit for making us see in the negative result of experiments like those of Michelson, Rayleigh and Brace, not a fortuitous compensation of opposing effects, but the manifestation of a general and fundamental principle.

Lorentz therefore acknowledges that Einstein's solution would be preferable, also because the transformations had been simplified with respect to the form he had found. Lorentz continues this way[47]:

> It would be unjust not to add that, besides the fascinating boldness of its starting point, Einstein's theory has another marked advantage over mine. Whereas I have not been able to obtain for the equations referred to moving axes exactly the same form as for those which apply to a stationary system, Einstein has accomplished this by means of a system of new variables slightly different from those which I have introduced. I have not availed myself of his substitutions, only because the formulae are rather complicated and look somewhat artificial, unless one deduces them from the principle of relativity itself.

Transformations in the original form of Lorentz's text are not as simple as in the form then consolidated, but this is not conceptually relevant: it would only be if one wanted to return to Lorentz's hypothesis. It is significant, however, that with this kind of consideration, the ground of a mentality inclined to accept Einstein's theory was prepared by closing eyes on the intrinsic inconsistency which makes it devoid of physical significance, in exchange for the advantages it would offer if it were internally consistent.

Before continuing, there is still one aspect of the problem that we must not underestimate: Poincaré had given a special methodological significance to the generalization of the law of inertia, extending it into the principle of relativity, as if the coexistence of the law of inertia with the absolute reference constituted by the ether were a kind of offense to the needs of rational science unity. This is a non-empirical attitude, and if you want, you could easily object to it: if

---

[47] Lorentz 1909 and 1916, p. 229-230.

we were to find experimental evidence beyond any doubt that the ether exists and at the same time that things continue in their rectilinear motion until a force does not lead to a change of direction, what else could we do else than accept the fact that nature is thus constituted? Nature generally has configurations that for us men are empirical matters of fact and are not explained by any rational principle, but simply exist. The need for rational unity of human thought has the right to criticize the logical errors of our thinking processes, but not to prescribe to nature that it should exist in the way that we would like better. However, Poincaré would say, the law of inertia is not a certainty, but it is the fruit of a convention which assumes the generalization of the empirical evidence that things continue on their rectilinear path until a force hinders them; but the fact that things really go on endlessly has never been experienced by anyone. And with regard to the ether, in any case, with or without Lorentz, it is an abstract entity constructed by hypothesis. Therefore, for a requirement of unity of thought, Poincaré had the goal to reconcile the law of mechanical inertia with the laws of electromagnetism without resorting to such a strong and specific hypothesis as Lorentz's. In the St. Louis lecture, as we have just seen, Poincaré had launched a sort of a call to arms to find a "more satisfactory" theory for electrodynamics of moving bodies, not knowing that he was suggesting to Einstein the subject and the title of the next year's essay. And given the immense prestige of Poincaré in those times, that call to arms offered the celebrity award to whom could answer it.

In 1909, Gilbert Lewis wrote a new paper on the relationship between energy and mass, in collaboration with Richard Tolman, who would later become a proud supporter of relativity. Both were very young, Lewis was 34 and Tolman 28. In this paper we find an exposition of known concepts of relativity developed in the usual way, through examples of systems in reciprocal motion and synchronization of clocks, and then the assertion[48]:

> Let us emphasize once more, that these changes in the units of time and length, as well as the changes in the units of mass, force, and energy which we are about to discuss, possess in a certain sense a purely

---

[48] Lewis and Tolman 1909, pp. 717-718.

factitious significance; although, as we shall show, this is equally true of other universally accepted physical conceptions. We are only justified in speaking of a body in motion when we have in mind some definite though arbitrarily chosen point as a point of rest. The distortion of a moving body is not a physical change in the body itself, but is a scientific fiction.

When Lorentz first advanced the idea that an electron, or in fact any moving body, is shortened in the line of its motion, he pictured a real distortion of the body in consequence of a real motion through a stationary ether, and his theory has aroused considerable discussion as to the nature of the forces which would be necessary to produce such a deformation. The point of view first advanced by Einstein, which we have here adopted, is radically different. Absolute motion has no significance. Imagine an electron and a number of observers moving in different directions with respect to it. To each observer, naively considering himself to be at rest, the electron will appear shortened in a different direction and by a different amount; but the physical condition of the electron obviously does not depend upon the state of mind of the observers.

Although these changes in the units of space and time appear in a certain sense psychological, we adopt them rather than abandon completely the fundamental conceptions of space, time, and velocity, upon which the science of physics now rests. At present there appears no other alternative.

As we can see, the interpretation of 1909 was clearly conventional: in fact, it was so conventional that it attributed "factitious" character also to mass gains that in truth Lewis had described as entirely real one year earlier. But this was exactly the problem: if the contraction phenomena "appear in a certain sense psychological", and the theory is a "scientific fiction", what relevance can it have for the description of objective physical facts? The same authors of the paper tell us that "the physical condition of the electron obviously does not depend upon the state of mind of the observers": but what we want to know are exactly the objective properties of things, regardless of the mental state of ourselves and every other observer. Above in section 3.14 we have examined in detail what would happen by adopting the theory in the conventional sense, and it is not necessary to repeat it; here the authors adhere to the conventional interpretation, but the consequences are not carried out and remain vague. As for the concept of "scientific fiction", we have seen that in the passage quoted the authors had said these very words[49]:

> Let us emphasize once more, that these changes in the units of time and length, as well as the changes in the units of mass, force, and energy which we are about to discuss, possess in a certain sense a purely factitious significance; although, as we shall show, this is equally true of other universally accepted physical conceptions.

And here the objection cannot be passed over in silence. Not only are there are other "universally accepted" physical concepts that are "purely factitious", but one can well say that all physical concepts are abstractions, because in any case phenomena are observed, they are measured, a model is formed that could correspond to the objective reality of the phenomenon and equations are written that allow us to predict the measurements of the phenomenon as a function of other ones. Conventions are everywhere, and conventions are even more relevant than the authors here say, not less. But the fictions, or abstractions, or conventions that are stipulated, are done with the aim of achieving unique results, while here we know from the beginning that the convention would take us to ambiguous measures. And so we know that Einstein's relativity is a very *sui generis* convention, which the authors of the paper show to know too, and which they try to justify showing some embarrassment, with the pre-established wish to find reasons to accept and maintain the picture outlined by Einstein. However, in the authors' words quoted above, we have read some that show that they realized what they were saying:

> Although these changes in the units of space and time appear in a certain sense psychological, we adopt them rather than abandon completely the fundamental conceptions of space, time, and velocity, upon which the science of physics now rests. At present there appears no other alternative.

The problem was to eliminate the exception to the law of inertia (the law of inertia is the "fundamental conception" here recalled) represented by the concept of ether and absolute motion over ether, and up to about 1909, it was attempted to achieve this by the ambiguous and unconvinced acceptance of the "fiction" and the

---

[49] Lewis and Tolman 1909, p. 717.

assumption that it could lead to results maybe still factitious but somewhat useful. This is what the authors tried to accomplish, and the sense of the attempt is clear, but getting a consistent theory is impossible. We understand what they would like to get, but there is no consistent and convincing description of the result.

As for the paper of Lewis and Tolman, it is worth mentioning a small detail that contains in a nutshell far more important consequences than itself. Above we have read:

> Imagine an electron and a number of observers moving in different directions with respect to it. To each observer, naively considering himself to be at rest, the electron will appear shortened in a different direction and by a different amount; but the physical condition of the electron obviously does not depend upon the state of mind of the observers.

The meaning of the argument is so obvious that it seems useless to comment on it: everything is in motion with respect to countless others at different speeds, and therefore having eliminated the unique reference of the ether, Einstein's postulate can only be interpreted in conventional or apparent terms, because in the case of realist interpretation, everything should undergo countless real different contractions at the same instant. Thus, according to the authors, the "physical condition of the electron" is one and "obviously" it does not depend on the mental states of the infinite observers in motion that may exist. Why do we spend time commenting on something so simple? Because this subject no longer appeared in relativistic literature. Soon began a debate about the possibility of realist interpretation, and discussions and attempts to demonstrate the possibility of realist interpretation would always focus on the artificial situation in which we compare two systems to try to find a reason to determine which between two clocks in motion will delay, not which among n clocks. The real-world situation, in which there are not two but n related systems in motion, were forgotten with a truly awesome removal process, and the consideration expressed by Lewis and Tolman in 1909 would no longer appear in the typical discussions that are read in relativistic literature. The paper of 1909 could afford to mention the problem because then the conventional or apparent interpretation was taken for granted and the authors tried to find the way to turn it into something useful, but the realist interpretation was not yet taken into account.

5.5   Transition from apparent to realist interpretation (1911)

The transition from the conventional or apparent interpretation of space and time dilatation to the realist one was required in order for the theory to have physical and not merely mathematical meaning: in fact, the transition took place and the realist interpretation became prevalent. It would be interesting to reconstruct the story exactly through a scrutiny of all the papers published in those years, but in this book we limit ourselves to identifying some fundamental moments and choosing some testimonies that we have reason to consider meaningful because they are Einstein's assertions, or because they correspond to events whose knowledge certainly was widespread in the audience of their era.

In a lecture in January 1911 in Zurich, Einstein dared to utter these words[50]:

> The thing is at its funniest when one imagines that the following is being done: One imparts to this clock a very great velocity (almost equal to c), then lets it fly on in uniform motion, and after the clock has covered a long stretch, one imparts to it a momentum in the opposite direction, so that it returns to the point from which it has been launched. It then turns out that the positions of the clock's hands have hardly changed during the clock's entire trip, while an identically constituted clock that remained at rest at the launching point during the entire time changed the setting of its hands quite substantially. It should be added that whatever holds for this clock, which we introduced as a simple representation of all physical phenomena, holds also for closed physical systems of any other constitution. Were we, for example, to place a living organism in a box and make it perform the same to-and-fro motion as the clock discussed above, it would be possible to have this organism return to its original starting point after an arbitrarily long flight having undergone an arbitrarily small change, while identically constituted organisms that remained at rest at the point of origin have long since given way to new generations.

> The long time spent on the trip represented only an instant for the moving organism if the motion occurred with approximately the velocity of light! This is an inevitable consequence of our fundamental

---

[50] Einstein January 1911.

principles, imposed on us by experience.

The text is well known because the Einstein lecture was held for the *Naturforschende Gesellschaft* Zürich, the association of researchers in Zurich, and was soon published. The Princeton University edition informs us that the text was revised by Einstein[51]. Not only does Einstein claim that the delay of a clock as a result of the motion state is real, but expressly develops the inevitable consequence that time dilatation implies the slowdown of all the processes taking place in the moving system, including the organic and biologic ones. And it is clear that this eliminates any ambiguity about conventional or apparent interpretation, which is excluded, because even if the synchronization of two clocks through remote procedures could give rise to some artificial differences between time indications, synchronization surely could not cause aging or rejuvenation of an organism.

What is said here by Einstein is perfectly equivalent to the renowned twin paradox, which first saw the light of the day in April 1911 when the fourth "International Congress of Philosophy" was held in Bologna, Italy, a meeting of philosophy scholars and scientists which had remarkable echo for the quantity and quality of the participants. At the meeting Paul Langevin read a conversational, brilliant, rhetorical, and exuberant report about the novelties of contemporary physics, and among other things he reported the new Einstein theory by making the example of two twins, one of which runs at a very high speed on a bullet, then returns to the Earth, and finds himself younger than his brother because his clocks and his body have undergone the contraction of the durations corresponding to the dilatation of time calculated according to Lorentz's formulas. In the example of Langevin, the astronaut's speed is determined so that Lorentz's transformations make the journey two hundred years long for the Earth and two years only in flight, on the whole return journey. Langevin shows an awareness of the fact that the symmetry of motion between the Earth and the astronaut, given the law of inertia, would not allow a determination of who is on the move and who is at rest between the two, and therefore would not let us know

---

[51] http://einsteinpapers.press.princeton.edu/ documento 17 del 1911

who is younger than the other. The narration is suggestive enough for a distracted listener to ignore this problem, but Langevin took his precautions, describing the phenomenon also as apparent, not without ambiguity[52]:

> ... it must be noted that the Earth will take two centuries to receive the signals sent by the explorer during the displacement movement that lasts just one year for him; it will see him live during this time a life slowed down two hundred times; it will see him perform the gestures that take place in a year.

When the traveller is on the way back, the Earth dwellers will see him much faster because the traveller will be anticipated by the signals he has emitted, and closing the accounts he will arrive at Earth aged two years against the two centuries spent on the Earth. Why does the traveller return young, and not the terrestrial twin (or rather, the whole Earth with all its organisms), since each one moved at the same speed with respect to the other? Because[53]:

> ... will be less aged those whose movement during the separation will have been more different from the uniform motion, those who will suffer more accelerations.

This statement shows that Langevin knows very well that the law of inertia does not allow the dilatation to depend on the condition of motion and speed, and therefore introduces acceleration. But the statement has a tremendous scope: not because Einstein did not speak of acceleration, but because the theory would be completely reformulated. If the cause of dilatation is acceleration, the formula applied for the determination of dilatation must be a function of acceleration, not speed. Langevin is saying that if a system moves from A to B at constant speed v having accelerated once on

---

[52] Langevin 1911, p. 51. "... il faut remarquer que la Terre mettra deux siècles à recevoir les signaux envoyés par l'explorateur pendant son mouvement d'éloignement qui pour lui dure un an: elle le verra vivre pendant ce temps dans son arche d'une vie deux cents fois ralentie; elle lui verra accomplir le gestes d'un an."

[53] Langevin 1911, pp. 49-50. "... ceux-là auront le moins vieilli dont le mouvement pendant la séparation aura été le plus éloigné d'être uniforme, qui auront subi le plus d'accélérations."

departure, once at half way to reverse the direction of travel and another time to stop, hence overall three times, and another system makes the same journey at the same average speed v, but stopping and restarting one thousand times, hence decelerating and accelerating overall two-thousand and three times, the second system should undergo a greater dilatation than the one that only suffered three accelerations. This is Langevin's idea, which says that whoever is undergoing more accelerations remains younger: but if this is the case, then the theory is to be rewritten as a function of the amount of acceleration, and all that is done is lost, including the result of Maxwell's equations invariance, because, as we know, to preserve that result it is necessary that instead of the Galilean transformations exactly those of Lorentz are applied, which are functions of speed. Other objections would be possible, but it is not worth it, because the context shows to anyone who can read that Langevin's words are inventive expedients to support a speech with which he wanted to astonish an audience, and nothing that has to do with the wording of a more consistent theory. Those who wish can read the whole Langevin conference online[54] and realize its celebratory and triumphal character.

The statement of the theory in the form of the tale of the twins proved a well-designed expedient. It was immediately said what is said still by anyone who hears that story: that it is impossible from the technical and biological point of view, but on a microscopic scale the phenomenon happens. The non-specialist public does not realize the problem of the incompatibility of the theory with the law of inertia, and understands even less how unjustified and imaginative it is to resort to the argument of acceleration, since the mathematics of the theory are conceived as functions of speed. This happened at the Congress in Bologna, and is still happening nowadays.

In May 1911 we find a page by Einstein that comes back to the subject, and shows how the dilemma between apparent and realist interpretation had come to mind for the specialist readers and demanded answers. The mathematician Vladimir Varičak had taken part in the discussion of a problem then known as *Ehrenfest's*

---

[54] See Bibliography.

paradox: the problem asked what would happen to a rotating disk subject to Lorentz's contraction, and the discussions formulated hypotheses about the various physical consequences that could occur. Varičak had noted that the problem is sensible in the frame of Lorentz, but not in that of Einstein, where nothing happens to the disk and contraction is only seen by the observer. So Varičak expressed himself in the sense of the conventional, or apparent interpretation, considering it an obvious thing, and Einstein answered with a note of two pages[55]:

> Recently V. Varičak published in this journal[56] some comments that should not go unanswered because they may cause confusion. The author unjustifiably perceived a difference between Lorentz's conception and mine with regard to the physical facts. The question of whether the Lorentz contraction does or does not exist in reality is misleading. It does not exist "in reality" inasmuch as it does not exist for a moving observer; but it does exist "in reality," i.e., in such a way that, in principle, it could be detected by physical means, for a noncomoving observer.

We see that the problem is so obvious that the answer must now resort to rhetorical artifices under which simple trivialities are hidden. Einstein's argument has become that contraction is "real" because it is "observable" with objective means, which is the same as saying that the image of a lean person seen through a deforming lens that makes him or her look fat is "real" because we can "observe" the relationship between the width of the person seen through the deforming lens and the width seen without it. That said, Einstein invokes the usual argument for which[57]:

> ...the Lorentz contraction has its roots solely in the arbitrary stipulations about the "manner of our clock regulation and length measurement."

This means, the argument that the theory of special relativity would define an objective measurement method, while the methods that have always been used would be based on arbitrary assumptions, "willkürlichen Festsetzungen". The "arbitrary assumptions" among others would have forced Lorentz to assume that contraction is a

---

[55] Einstein May 1911.

[56] See Varičak 1911.

[57] Einstein May 1911.

physical fact manifested in particular circumstances, while it is a universal mathematical fact.

Given the shortness of the paper, in it the elements of the theory developed elsewhere are invoked without explicitly mentioning, but Einstein's argument is more precisely this: traditionally, no attention was paid to the fact that remote clock synchronization requires the definition of appropriate procedures, and requires the assumption that the speed of light is constant in order to ensure that the signals' journey forward and the journey to come back are equal or comparable. The traditional disregard for this is the "arbitrary assumption" that has always trivialized the problem of synchronization.

This says Einstein, and the relief is correct, but it has nothing to do with either the question of the interpretation of the theory or with Einstein's specific postulate. To synchronize two remote clocks, we must assume that the speed of light is constant either with respect to the ether or the emitting body, and then act accordingly to the ether model or the ballistic one, but we need not assume that c is constant for each observer also in the ballistic framework, which is Einstein's specific postulate, and with which it becomes impossible to describe the kinematic behaviour of light except through the generic words that state the postulate itself. Einstein's specific postulate is not necessary to synchronize remote clocks, but only to preserve the invariance of Maxwell's equations with an act of wishful thinking. So in these lines one does not read the answer to Varičak at all, but called into question are other aspects of the theory that serve to evade the problem.

At the end we find a magnificent exemplar of *petitio principii*, a perfect example corresponding to the definition of this typical logical error, which is to invoke as a premise the conclusion that we want to demonstrate. Einstein, unable to answer Varičak, tries to prove once more that the theory of special relativity has no alternative. The old non-relativistic technique of measuring space and time is a "point of view that cannot be preserved" and hence[58]:

The following thought experiment shows to what extent this view [the

---

[58] Einstein May 1911.

non-relativistic technique of measurement] cannot be maintained. Consider two equally long rods (when compared at rest) A'B' and A"B", which can slide along the X-axis of a non-accelerated coordinate system in the same direction as and parallel to the X-axis. Let A'B' and A"B" glide past each other with an arbitrarily large, constant speed, with A'B' moving in the positive, and A"B" in the negative direction of the X-axis. Let the endpoints A' and A" meet at a point A* on the X-axis, while the endpoints B' and B" meet at a point B*. According to the theory of relativity, the distance A*B* will then be smaller than the length of either of the two rods A'B' and A"B", which fact can be established with the aid of one of the rods, by laying it along the stretch A*B* while it is in the state of rest.

This is the corresponding figure:

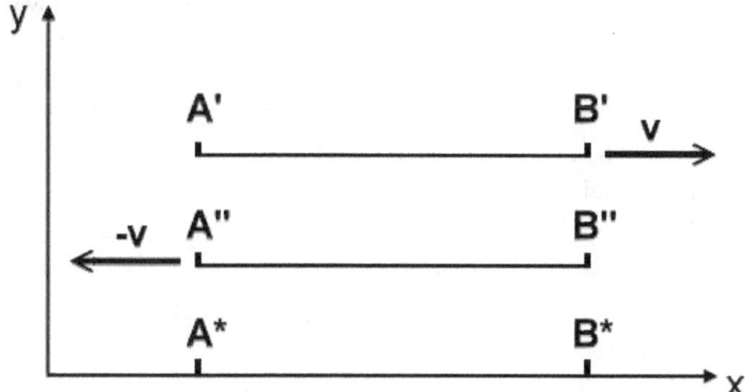

The ruler A'B' having length l moves at speed v, while the ruler A"B", which has the same length l, moves at speed $-v$ with respect to the X-axis. When the extremes coincide, if we measured with light signals the distance between the points A* and B* on the X-axis, we would find a measured distance l' which would be shorter than l. And if, later, to check at rest, we used one of the two rulers overlaying it to the segment A*B* marked when the two rulers were in motion, we would verify that l' is shorter than l.

So, besides confirming that the distance l' is apparent (it is Einstein who is saying that once at rest of the rulers would measure the distance l and not l', and that the segment A*B* is an image formed by measuring distance while in motion), the sense of the argument is: for what reason must we adopt the relativity theory, and deduce that a given length is different from what we would expect without the relativity theory? For the reason that the relativity theory asserts that a given length is not what we would guess without the relativity theory. This is, therefore, the argument: the theory is right because according to the theory the theory is right.

This is what we can accomplish reading Einstein's paper, who in 1911 cannot escape the problem of incompatibility between the need for realist interpretation of dilatations and the law of inertia that can only admit the apparent interpretation. The argument used is fully a *petitio principii*, coarse and brazen. The expression *petitio principii* is generally erroneously phrased "begging the question", while it should be translated "assuming the point to demonstrate". This is the definition of *petitio principii* according to a basic manual of logic[59]:

> *Petitio principii*. It consists of assuming the truth of just what one wants to prove as the basis of the demonstration. The authors of Port-Royal Logic claim as an example the reasoning by which, according to Galileo, Aristotle wants to prove that the Earth is at the centre of the Universe:
>
>> "The nature of heavy things is to tend towards the centre of the Universe and of light things to distance from it;
>>
>> But experience shows us that heavy things tend to the centre of the Earth and that light things go away;
>>
>> Therefore the centre of the Earth coincides with the centre of the Universe."
>
> The *petitio principii* is in the major premise, because only assuming the conclusion (that the centre of the Earth is also the centre of the Universe) can be said that heavy things tend to the centre of the Universe, since experience only tells us that they tend to the centre of the Earth.

Petitio principii is one of the most common logical errors encountered in the relativistic literature. We shall see below how the arguments used in the relativistic literature with respect to the symmetry of spatial and temporal dilatation are structured, and how actual logical errors are usually mixed with rhetorical gimmicks, while in the text of Einstein we happen to encounter them expressed with careless innocence, as in the case just seen.

---

[59] Pesce e Pozzi 1971, p. 48.

5.6 *Realist interpretation of dilatation through general relativity (1918)*

Einstein rarely came back to the subject and only for hints, but in 1918 he was forced to publish a paper in which the problem was addressed by referring to it through the theory of general relativity, which as is known was published in the years 1915-1916. After the publication of general relativity, there was a strong scientific controversy regarding Einstein, and one of the most authoritative critics of relativity was the Austrian physicist Philipp Eduard Anton von Lenard, Nobel Prize in 1905, who despite being Austro-Hungarian had always been a philo-German nationalist, and who already in the early 1920s adhered to Nazism and then compromised very seriously on the criminal follies of the so-called "Aryan science". The controversy then immediately acquired an obvious political meaning, but I do not doubt that the readers of the following pages will be able to abstract from the political dimension and only evaluate the meaning of Einstein's words for the scientific problem that interests us.

The 1918 paper is titled "Dialogue about Objections to the Theory of Relativity" and has the form of a conversation in which an opponent of relativity called Kritikus poses objections to a character called Relativist who responds until he forces Kritikus to surrender. The dialogue wants to evoke, of course, that of Galileo, Kritikus plays the part of Simplicio and Relativist is Einstein himself. Lenard is mentioned explicitly several times, and it is clear that Kritikus is his spokesperson. The paper extends to some pages and concerns the symmetry problem: given two clocks in relative motion U1 and U2, which are in two systems called as always K and K', what is the criterion to determine which of them is delayed? Although this is the central problem, the words with which Relativist tries to answer the question specifically are very few, and we shall also quote them in German to pay attention to the nuances of meaning. For the rest, the paper digresses in considerations of general scope.

The paper opens with some considerations in which Kritikus, who as Galileo's Simplicio represents the academic corporation, declares the intention to express an honest criticism of relativity and without exerting the bad habit of professorial arrogance. Through this expediency Relativist paints himself as the weak part, and it is not said that in 1918 this corresponded to the truth, because the wind of public consent had long been favourable to Einstein rather than to the traditionalists: the 1918 victimhood sounds more artificial that sincere, and readers should not let themselves be misled by subsequent political facts in evaluating it. Kritikus goes into describing the problem where we have two synchronized clocks in the same place: if one of the two clocks moves along a straight line, then reverses the sense of motion and finally rejoins with the first, how can we decide which of the two clocks is delayed depending on relative speed, that can be attributed indifferently to each of the two clocks?

To introduce the problem Kritikus says[60]:

Ist aber eine der Uhren, z. B. U2, relativ zu K im Zustande gleichförmiger Translationsbewegung, so soll sie nach der speziellen Relativitätstheorie — vom Koordinatensystem K aus beurteilt — langsamer gehen als die relativ zu K ruhend angeordnete Uhr U1.

But if one of the clocks, e. g. U2, is in a state of uniform translatory motion relative to K, then it shall, according to the special theory of relativity—judged from the coordinate system K—go at a slower rate than the clock U1 which is still placed at rest relative to K.

Thus, in the initial definition of the problem, the apparent interpretation is taken for granted: according to theory any of the clocks is delayed when "vom Koordinatensystem K aus beurteilt", "judged by the coordinate system K", or in the measurements carried out by the system considering itself at rest. This hint to apparent interpretation is in the words of Kritikus, but the relativist does not care to refute it, and even declares himself to agree with the presentation of the problem as outlined by Kritikus, but to have other elements to consider[61]:

Es ist sicherlich richtig, daß wir uns vom Standpunkt der allgemeinen Relativitätstheorie aus ebensogut des Koordinatensystems K' bedienen können als des Koordinatensystems K. Aber man sieht leicht ein, daß die Systeme K und K' mit Bezug auf den betrachteten Vorgang keineswegs gleichwertig sind. Während nämlich der Vorgang von System K aus wie

---

[60] Einstein 1918.

[61] Einstein 1918.

oben aufzufassen ist, gestaltet er sich von K' aus betrachtet, völlig verschieden, wie die nachfolgende Gegenüberstellung zeigt.

It is certainly correct that in the general theory of relativity we can use the coordinate system K' as well as the coordinate system K. But it is easily seen that the systems K and K' are by no means equivalent when it comes to the process under consideration. Although the process, when seen from system K, is to be interpreted in the manner outlined above, when viewed from K' the process is something completely different, as shall be demonstrated in the juxtaposition as shown below and in the figure.

At this point, the relativist constructs a representation of the elements at stake invoking concepts of general relativity. Unlike relativistic literature, it seems that Einstein never mentions acceleration within the framework of special relativity, and thus avoids exposing himself to the obvious criticism that would object that dilatation should then be considered as a function of acceleration rather than speed. One should read all published papers to see if Einstein really never mentioned acceleration directly as a factor of special relativity (unlike other authors, as we shall see later); however, in this case, he uses a more convoluted construct through general relativity. According to general relativity, which is a theory of both gravity and accelerated motion, accelerated motions are always reciprocal to gravitational phenomena: when one thing accelerates with respect to another, the other "drops" in the opposite direction. According to the theory, accelerations and gravity determine time contractions and clock frequency changes for specific reasons, different from those of special relativity. Therefore, the argument of the relativist becomes: the delaying clock is U2 because U2 undergoes three accelerations: one to start, one to invert the motion sense and one to stop when it returns to where U1 is located. But the net result is not that U2 is delaying, but it is that U1 anticipates, because when U2 is accelerated U1 is affected by reciprocal gravitational phenomena correlative to U2 accelerations, and for this reason U1 undergoes an increase in its frequency. In brief, the argument is: when U2 accelerates to the right, U1 mutually "drops" to the left, because the acceleration affecting U2 determines the birth of a gravitational field that affects U1. When U2 accelerates to the right, U1 "drops" to the left for the same reason. As a net result, the U1 frequency increase during the three U2 acceleration phases is greater than the slowdown of not only U2, but also U1, and then U1 at the end indicates a greater time value than U2.

The process is described in five phases, which represent the same physical phenomenon from the point of view of K and K':

	Frame of K	*Frame of K'*
1.	acceleration of U2 towards the right until speed v is reached	"it comes into existence" (the German expression is "es entsteht") a gravitational field that impresses to U1 speed $-v$ towards the left
2.	motion of U2 at constant speed v	motion of U1 at constant speed $-v$
3.	acceleration of U2 to invert the sense of motion	"it comes into existence" ("es entsteht") a gravitational field to invert the speed of U1
4.	Motion of U2 at constant speed $-v$	motion of U1 at constant speed v
5.	deceleration of U2 until speed zero	"it comes into existence" ("es entsteht") a gravitational field to take to zero also the speed of U1

And this is the conclusion[62]:

Es ist wohl im Auge zu behalten, daß in der linken und in der rechten Spalte genau der nämliche Vorgang beschrieben ist, nur bezieht sich die Beschreibung links auf das Koordinatensystem K, die Beschreibung rechts auf das Koordinatensystem K'. Gemäß beiden Beschreibungen ist die Uhr U2 am Ende des betrachteten Prozesses gegenüber der Uhr U1 um einen bestimmten Betrag zurückgeblieben. Bei Beziehung auf das Koordinatensystem K' erklärt sich dies Verhalten folgendermaßen:- Während der Teilprozesse 2 und 4 geht zwar die mit der Geschwindigkeit v bewegte Uhr U1 langsamer als die ruhende Uhr U2. Aber dies Zurückbleiben wird überkompensiert durch einen schnelleren Gang von U1 während des Teilprozesses 3. Nach der allgemeinen Relativitätstheorie geht nämlich eine Uhr desto schneller, je höher das Gravitations-Potential an dem Orte ist, an dem sie sich befindet, und es befindet sich während des Teilprozesses 3 U2 tatsächlich an einem Orte

---

[62] Einstein 1918.

höheren Gravitations-Potentials als U1. Die Rechnung ergibt, daß dies Vorauseilen gerade doppelt so viel ausmacht, als das Zurückbleiben während der Teilprozesse 2 und 4. Durch diese Betrachtung wird das von dir angeführte Paradoxon vollständig aufgeklärt.

One must carefully keep in focus that both columns, left and right, describe exactly the same process, but the description on the left refers to the coordinate system K, the one on the right to the coordinate system K'. According to both descriptions, it is the clock U2 which lags by a certain amount behind clock U1 at the end of the process considered. With respect to the coordinate system K' the phenomenon is explained in the following manner: During procedural steps 2 and 4, clock U1 moving at speed v, has indeed a slower rate than clock U2 which is at rest. But the time lag gets overcompensated by the faster rate of U1 during procedural step 3. Because, according to the general theory of relativity, a clock has a more accelerated rate the higher the gravitational potential is at the clock's location; and during procedural step 3, U2 is indeed at a location of higher gravitational potential than U1. Calculation shows that this running-ahead amounts to precisely twice as much as the lag-behind during the procedural steps 2 and 4. This analysis clarifies completely the paradox you referred to.

For the sake of accuracy, let us note that there is a print error in the text so at the end U2 and U1 are exchanged: given the context, U1 is the clock that is in a place having higher gravitational potential because it "falls" while U2 accelerates[63]. But let us see what is the exact meaning of these lines. We read: "With respect to the coordinate system K' (...) during procedural steps 2 and 4, clock U1 moving at speed v, has indeed a slower rate than clock U2 which is at rest. But the time lag gets overcompensated (überkompensiert) by the faster rate of U1 during procedural step 3." So we have read that the clock U1 has a smaller frequency from point of view of K' in a reciprocal way U2 has with respect to K, so the phenomenon is apparent. But this lesser frequency needs to be compensated, and one does not understand why, if it is apparent. It is abundantly compensated by the effect of the gravitational fields that involve U1, and so in the end we find not U2 backwards than U1, but U1 ahead of U2. But is the frequency increase due to the gravitational field in

---

[63] The error is in the original printed text, as can be verified in http://einsteinpapers.press.princeton.edu

turn apparent or real? It "comes into existence", "es entsteht", when looking at things from the point of view of K', says the text, and therefore even the phenomenon determined by the gravitational field could be apparent: yet it must be real, because at the end of the process the two clocks do not mark the same time when they are compared locally.

Moreover, how do we want to judge the expression we just read, "... calculation shows that this running-ahead amounts to precisely twice as much as the lag-behind"? Are the effects of general relativity always twice as high as those of special relativity? If so, it means that we have only one theory, not two, and that the theory of special relativity alone no longer holds true for Einstein himself. Or is it meant that calculations compensate the phenomenon in this single case? Then with what right can we generalize the conclusion? Do we want to think that Providence will always compensate for the effects of special relativity with opposite and double effects of the general one?

I think it is necessary to interrupt the account with this consideration: one can go crazy if one reads these pages with the positive prejudice that the theories of special and general relativity are both serious and authoritative, given the general consensus they have received, and consequently if one tries to attribute a consistent meaning to these topics. But if we realize that everything was born from the attempt to preserve Maxwell's laws through a convention that would have to save the labour of studying the behaviour of electromagnetic waves in the ballistic hypothesis, then we realize that these speeches are only an attempt to reconcile the impossible, given the failure of the attempt, with arguments that make ridiculous both the people who write them and those who take them seriously. It was right to consider the 1905 hypothesis worth of study, as it proposed a simpler and more elegant solution than any other for the underlying problems. But the reason why specialists of that generation were obstinate to maintain this theory at the cost of building castles in the air of artificial and merely verbal solutions was increasingly complicated, and why later the world has given general consensus to these theories, this is not a problem of physics, but of anthropology, and we shall try to get a first idea about it in the following.

5. Time and space dilatation and law of inertia

Let us go back to the 1918 paper. Kritikus could now bring at least two issues into questions, and could ask:

1. If the reason for the real clock frequency change lies in the complex represented by acceleration and gravitational phenomena, why is the frequency variation calculated with Lorentz formulas as a function of speed? Does the providential case that U1 actually anticipates the exact amount of U2 apparent delay always happen? And if U2 instead of undergoing three accelerations underwent three hundred or three thousand while travelling overall at average speed v, would the three hundred or three thousand accelerations give rise to the same result as the three of the example? The average speed v would always be the parameter on which to base the calculation of the phenomenon, regardless of the overall force exerted on the U2 clock, and regardless of the overall "coming into existence" (or how you want to interpret "es entsteht") of gravitational fields?

2. And if there were three clocks, and two of these U2 and U3 separated from U1 by receiving different accelerations and at speed v_2 and v_3 respectively, then would U1 anticipate as a function of v_2 or v_3?

Instead Kritikus, who is really a simple and pious soul, merely asks[64]:

Du hast das Paradoxon gelöst, indem du den Einfluß eines relativ zu K' herrschenden Gravitationsfeldes auf die Uhren in Rechnung zogst. Ist aber dieses Gravitationsfeld nicht etwas bloß Fingiertes? Seine Existenz wird doch nur durch die Koordinatenwahl vorgetäuscht. Wirkliche Gravitationsfelder sind doch stets durch Massen erzeugt, und können nicht durch geeignete Koordinatenwahl zum Verschwinden gebracht werden. Wie sollte man glauben können, daß ein bloß fingiertes Feld auf den Gang von Uhren einen Einfluß haben könnte?

You solved the paradox by taking into account the influence on clocks of the gravitational field, which is relative to K'. But isn't this gravitational field only fictitious? Its existence is, I should say, only simulated by the choice of coordinates. After all, real gravitational fields are always generated by masses and cannot be made to vanish by a suitable choice of coordinates. How should one believe that a merely fictitious field could influence the rate of clocks?

---

[64] Einstein 1918.

5. Time and space dilatation and law of inertia

At this point we have not even reached half the length of the paper, but about the specific problem to which we had to answer there is no longer any word. Kritikus could ask sharp questions and solicit unambiguous answers, and instead he ventures on the terrain of concepts of reality and appearance related to the gravitational fields invoked by the relativist, and thus he triggers an interesting lesson about methodology and philosophy by the relativist. The relativist moves from what is found and has always been found conventional in classical mechanics, he blames the naivety of Kritikus for a long time and of course speaking of general things he expresses plausible concepts making a very good impression. Only, the relativist's methodological principles neither respond to the initial question nor show how could be plausible the imaginative solution for the problem that had been proposed, which is obviously fallacious and depends on the benevolence of Divine Providence that makes special and general relativity calculations agree.

Finally, the reader may wonder: but if the solution to paradoxes is in general relativity, why is the special one not abandoned? The question is legitimate, but the answer is simple: general relativity presupposes special relativity, and it presupposes not only time dilatation in uniform speed motion but also the indispensable Lorentz transformations, which in general relativity develop into much more complex mathematical elaborations, but remain the basis of everything. But above all, general relativity presupposes and exploits the same rhetorical artefact born (probably by chance) by writing special relativity: the always-unsolved ambiguity between presenting a phenomenon both as apparent and real and the ability to exploit the indeterminacy of speech and brazen contradictions to create the illusion of a superior depth of thought.

6.   *Discussions of relativity paradoxes*

6.1   *After Einstein*

The relativistic literature is an immense collection of contradictions, developed over a long time and thousands of pages, but adhering to a few recurrent patterns and rhetorical artifices. After reading some sources, one finds the same expressions and narratives, with slight variations over the ages. The updated papers of the last few years shed no light on the ones of decades ago. The central problem is to maintain a somewhat plausible picture of the theory without deciding between the conventional or apparent interpretation and the realist interpretation, and hence to conceal and deny the problem of the incompatibility of the theory with the law of inertia.

Now we shall see some paradigmatic examples of how this happens; later we shall see why this has happened and continues to happen, or at least we shall try to shape a first idea of the cultural and anthropological aspect of the problem. But soon we must consider a factor about which we shall fully persuade ourselves examining the motivations of belief in relativity: we are faced with a spontaneous phenomenon whose actors are unable to manage the excessive complexity of the problem. There are no conspiracies, and those who write the absurdities that are read in relativistic literature have interiorized the contradictory expression they repeat because they always have a reason to do so. In the original case, the reason is that by eliminating relativity we would not find an easy alternative, but we should admit an ignorance of fundamental phenomena, and we would need to reformulate radically some points of view of physics. But this is the kind of job that belongs to people of the hard temper of Galileo and Newton, and not to the authors of books seeking arguments to save Einstein's immature hypotheses instead of discussing them critically. In the less noble case, the reason is very easy to understand: by interiorizing the contradictions with which a person can flaunt having understood relativity, one takes part in the banquet of the social elite of theoretical physicists, to which everyone would have the ambition to be invited.

To accept a contradiction, and here we are fully in the case that any people who accept a contradiction first deceive themselves and therefore then are sincere with others, one must exploit the inability of human memory to have a state of consciousness equally attentive to every element that make up an articulated problem. If a person contradicts him or herself in two lines, then all perceive contradiction, and even he or she who asserts it realizes it, unless the emotional reasons that lead him or her to contradiction are so strong that the person is immersed completely in the role he or she is playing.

Instead, if the development of a contradiction lasts ten or one hundred pages, the error becomes invisible to all or almost all; only those who carefully control the logical consistency of the proposed solutions for a problem, and break down into their elements the assertions they read with analytic criterion and with an indifferent attitude to the commonplace that all repeat, remain sheltered from inconsistency. So to hide the contradiction we find in relativistic literature unnecessary complications of the mathematical level, and we find that the abstraction level of the expression is raised without any reason. In this way the fact is exploited (for deceiving and self-deceiving) that few have a strong understanding of the relationship of mathematics with physics. Physics observes phenomena, measures them, uses imagination to invent schematic patterns that may correspond to the structure of an observed phenomenon, and finally writes equations that allow us to predict the measurements of the phenomenon as functions of others. Since at the end of the game calculations are made, the result of which is to obtain information on what happens and will happen, many believe, including specialists, that physics can demonstrate the reality of things deductively, as well as mathematics demonstrates its theorems. And so is exploited the illusion that mathematical consequences derived correctly from their premise can demonstrate the premises themselves. At the end of the whole process, those who have assimilated the theory stop asking questions about the mental path through which the theory is learnt; the habit becomes that of applying the Lorentz gamma factor to numerous theoretical problems devoid of technical and empirical verification without ever doubting the logical legitimacy of what is being done, thus repeating the operation performed for the first time by Einstein when he opened Chapter 6 of ED1905 by applying to the measurements of certain electromagnetic phenomena the transformations he had previously described as a result of a convention that gave rise to the apparent slowdown of a clock adjusted specifically so as to delay.

This is the picture, and anyone who has tried to study the theory of relativity with commitment, but with no satisfaction, has experienced it. All those who have tried to study relativity by deepening the premises instead have experienced that understanding the meaning of the M+M experiment is within reach of anybody, and that the theory of contraction of FitzGerald and Lorentz is also at the reach of anybody; to everyone is perfectly clear why in the hypothesis of Fitzgerald and Lorentz travellers who move with respect to the ether at high speed would experience the dilatation of time and the contraction of the durations of phenomena in the system in which they travel. Everyone has experienced that understanding these concepts in the old, arbitrary but coherent hypothesis does not require specialized knowledge, but just careful reading. In the version of Einstein, however, it is impossible to overcome the incompatibility of theory with the law of inertia and the supposed understanding consists of self-education to accept, in exchange for emotional gratification, the verbal expression of some mutually exclusive propositions to which no physical meaning corresponds, because contradiction is never in things, but only in human expression. It goes without saying that those who genuinely confess that they have never understood relativity have good reasons to estimate their own intelligence and their ability to resist suggestions, and much more if they insisted on trying to seize the theory without success.

It would not be useful now to make long or short quotations to show that the general stylistic imprint of the relativistic literature is this omnipresent attempt to make contradictory assertions acceptable, because this cannot be effectively demonstrated by isolated examples. One must live the experience of reading the relativistic textbooks, to know how within them contradiction pervades everything, and how authors try to make the reader complicit with them in accepting and hiding contradictions by mixing enthusiastic enticements and threats. There is a typical and easily recognizable rhetoric pervading all relativistic literature: the emphatic tone depicting the beauty of theory alternates with the premodern naivety qualification attributed as punishment to those who do not understand it. Often the textbooks that explain relativity describe its background and assumptions, and in doing so, they describe accurately numerous topics of classical physics. As long as they speak of these things, the expression is precise and the authors master what they say, so that while some relativistic treatises are clearly mere compilations, others are useful and interesting readings before they begin to talk about relativity, and sometimes the authors even seem to have personality in the description of classical themes; then, as soon as the speech comes to the main subject, the authors change tone and the suggestion of images alternates with non-conclusive reasoning with which the authors attempt to persuade both themselves and the readers that they own a consistent theory. For anyone who has experienced this while trying to understand relativity, all this has become self-evident, while those who do not have this experience must live it on their own if they want to verify the accuracy of our description of the situation.

Relativistic literature always dedicates a long central part of its treatises to the justification of the application of Lorentz's formulas to relative speed, taking for granted the apparent character of the phenomenon of temporal dilatation. This is done with arguments derived from Einstein's seminal one, which we have analysed word by word, and in this part authors always keep the ambiguity about the conventional or apparent or realist interpretation of what is described, but the prevailing suggestion is that of apparent interpretation: clocks in motion do not undergo any change, and time dilatation is perfectly symmetrical between two systems and is just a point of view of each observer. In the volume dedicated to relativity by the prestigious Routledge Guides series we read a funny example of readers' flattery. The concept of time dilatation was introduced by the method of the light-clock arguing it as described above in section 3.8, so each observer sees the light-clock of the other describing a sawtooth trajectory, and applying the second postulate of Einstein both observers attribute a smaller frequency to the moving clock and derive the notion of dilatation[65]:

> Before proceeding with the question of time dilatation, let's ask a
> question to make sure we understand where we are. Given that the

---

[65] Trefil 2015, Chapter 6.

observer on the ground observes the moving clock slowing down, what does the observer on the train see? Is the ground clock ticking faster or slower than the one on the train? If you answered 'slower', you've understood the argument. If you didn't, convince yourself that the observer on the train observes his or her clock registering the proper time and sees the light in the ground traveling in a sawtooth on the left.

Later, and we shall quote the words below, the same book and the same author will say that the GPS system clocks are subject to temporal dilatation in a non-apparent way. This is generally the case: usually in chapters presented as appendices or in depth discussion or pretended experimental evidence of relativity, the literature decides for realist interpretation, which is the only one that can be useful in some way, and thus asserts that time dilates as a function of speed, which determines the real slowdown, local and not symmetrical, of clocks and phenomena of all kinds. Realist interpretation is then propped up with some typical and recurring arguments in which the contradiction that pervades everything is manifested with immense evidence to the reader who has not been intimidated by the previous rhetoric. We shall now discuss in detail some of the typical arguments with which relativistic literature seeks to consolidate realist interpretation. All this continues to be referred only to special relativity: general relativity is almost never called into question, and the argument developed by Einstein in 1918 was only followed by few authors.

6.2   The μ meson

The discussion of the particle called μ-meson contains in a few lines all the problems of special relativity. The fact that this discussion is almost identical in all the treatises of the last decades, written after the discovery of this particle, makes the discussion deserving of quotation, and transforms it into an integral part of the theory.

We shall proceed in degrees: first we shall see in a macroscopic example how the argument is structured, and the readers will find no difficulty in judging it with their own intelligence. But the first example will only be analogous to the meson argument in the literature, and for this reason as a second step we shall expose these arguments by simplifying the numbers. Finally, two authors will be quoted literally, and the readers will judge the arguments by their own.

The meson argument is structured so as to maintain the symmetry between two systems in motion, thus declaring apparent the measurements made by applying the Lorentz formulas, and then deriving non-apparent consequences from these measurements. The example that simplifies the subject is this: let us imagine a man travelling quickly in a car to a destination that he must reach by 20:00 that day, because at that hour a deadline for a certain deal will expire. He looks at his clock and calculates that travelling all day long at the maximum speed of the vehicle, he will be able to reach the destination at 20:01. So he despairs, but it comes to mind to him that in his car there is a powerful laser telemeter capable of measuring distances of hundreds of kilometres. He then points it to the destination, reads the distance, considers his speed, recalls Einstein's second postulate, applies the contraction factor according to Lorentz and finds that the distance has diminished enough to allow him to arrive at 19:59. Or perhaps the telemeter returns the reduced distance by itself, because the phenomenon of contraction in the measurements happens spontaneously, who understood exactly how the remote measures would take place given the relativity concepts? So anyway the driver deduces that he will arrive a minute before the deadline.

Is the conclusion correct? Let us recall that the theory has taught us that if we are in motion with respect to an external system and we perform measurements of it using electromagnetic waves, for example using a laser telemeter, and if we make calculations according to ballistic theory, because it is postulated that "the speed of light is the same for every observer", the resulting measurements should be corrected with the Lorentz formulas we know. But according to the theory it is the measure made in this way that it is contracted: if we measure the soil with rulers there is no contraction, not because Einstein said this (which he also said), but because we are not in the world of Lorentz and we are not moving with respect to the ether that results in a real contraction. So if the car has to travel, say, two thousand kilometres with its wheels that have a circumference of two metres, it will need one million spins of its wheels and no less to get to the destination, and it will arrive at 20:01 as predicted by calculations made in the old way and not before. The μ meson argument is developed exactly like this: it describes a problem that would be resolved if a certain path could be travelled in dilated time or in contracted length as determined by remote measures, and suddenly it is stated that a path takes place in the dilated or contracted environment rather than in the proper one of the object in motion. That is, while the driver does not receive discounts by measuring the distance with a laser gun, the meson, according to the literature, does receive.

The μ meson, also called muon, is a particle whose behaviour shows a singular anomaly in experimental observations; this behaviour was interpreted through special relativity, considering the observed phenomenon a proof for special relativity. The proof, to call it so, is based on the obvious equivocation we have just exemplified and that is repeated in similar terms in all the textbooks that invoke the muon argument, but with different simplifications according to the choice of the different authors. We are now simplifying the experimental data and the structure of the argument that is considered a proof of special relativity, and then we shall read the story as it appears in some versions that refer exactly to the experimental data. We shall therefore verify that the developed argument corresponds exactly to our simplified scheme.

The experimental phenomenon observed is this: the muon is an unstable particle that can be produced in laboratory at much lower speeds than c, for which the Lorentz gamma factor is still without importance. In these conditions it is noted that muons live a certain time T (of the order of one millionth of a second), and then decay. The muons are also produced when cosmic beams penetrate the upper layers of the Earth's atmosphere. In this case, the muons travel at a speed V few smaller than c, which however is not enough to travel the thickness of the Earth's atmosphere, which we call l and which is about a few dozen kilometres, in their lifetime T. The muons should live a time long about 10 times T to cross the atmosphere and reach the Earth's surface where they are observed.

The observed data is therefore that the decay of muons takes place with a certain duration in laboratory conditions, and with a longer duration when the muons interact with the Earth atmosphere at high-speed. This is simply a physical fact, unexplained, about which one could look for a theory. However, this experience is interpreted in the light of special relativity in this way:

Premises:

- the proper time of muons' life, measured in a laboratory at non-relativistic speeds and with a gamma factor just a little more than 1, is T;

- the time needed to cross the Earth's atmosphere at speed V would be about T×10;

- so it is not understood how the muons can reach the Earth's surface;

- however, the muon's speed V in the atmosphere is high and corresponds to a gamma factor of about 10.

Let us try to derive consequences:

- It would be nice to say that a moving muon at speed V undergoes a factor 10 time dilatation, and hence a factor 0.1 contraction of its durations. In this case, the duration of all the phenomena of the muon's system should be multiplied by 0.1, decay would occur in ten times the time T needed at rest, and time T would be sufficient to reach the Earth. The theory of special relativity would explain the muon phenomena, and it would get an experimental confirmation.

- But we cannot state this because the phenomenon is perfectly symmetrical: the law of inertia does not allow us to say that it is the muon which is subject to time dilatation rather than the Earth. We might try to check what would happen by saying that it is the Earth that undergoes time dilatation and lengths contraction, and even in this case the time T would be enough, because the Earth's atmosphere would become about one tenth of l. But given the law of inertia, we have no right to say that the Earth undergoes dilatation and the muon does not. In addition, if we said this, everyone would notice that it is rather bizarre to think that the Earth at any instant is involved with a certain strong contraction factor with respect to the muons that come to it, and also infinite other factors with respect to the infinite celestial bodies with respect to which it is in motion.

- The last consideration also applies to muons, which are in motion at even relativistic speeds with respect to infinite different particles and entities in the Universe. Why should a muon contract by a factor corresponding to its speed with respect to the Earth, and not the one corresponding to the motion with respect to anything else?

Given the many reasons that prevent realist interpretation, the muon case is described as irreconcilable with the realist interpretation of the theory, and the authors do not invoke it, but stick to apparent interpretation. Nobody ever tries to say that the muons contract and the Earth does not, because the symmetry of the case is perfect. If a physical reason were invoked to hypothesize the slowdown of muons' decay, then we would abandon Einstein's relativity and return to classical physics, and precisely to some variation of Lorentz's theory. So since we cannot assert that it is only the muon that contracts and not the Earth:

- We say that the Earth-system moving at speed V measured remotely by the muon is time dilated with factor 10, and contracted in durations with factor 0.1.

- Mutually, we say that the muon-system moving at speed V measured remotely by the Earth is time dilated with factor 10, and contracted in durations with factor 0.1.

Having excluded realist interpretation, we remain in the conventional or apparent interpretation, which is perfectly symmetrical. So if the muon had conscious thinking and instruments with which to measure the Earth, it would find it small and slow. What follows? Nothing: the muon has to travel the distance l in the time T that is given it to live when it would need T×10. In fact, when a muon travels the distance l, it is in the condition that it can measure it with a ruler step by step, so the distance l does not change. The modified distance equal to l×0.1 is the one the muon would form in its brain if it measured the Earth's atmosphere by means of synchronized clocks and used a certain technique, that is, if it performed calculations according to simple ballistic theory. The modified distance l×0.1 is conventional or apparent, because the law of inertia does not allow realist interpretation.

At this point, it becomes clear what we know: the conventional or apparent interpretation is devoid of physical consequences. The simplicity of the case highlights in particular the uselessness of the conventional or apparent interpretation. How can we now state that the muon phenomenon is a proof of relativity?

- Let us write a sentence shaped like the following: "Since the muon's life T had been measured in the proper reference system of muons, we need to transform time T×10 into the proper reference system..." (*beware: for us this sentence is intentionally nonsense*).

Hence the following syllogism is developed:

- The muon's proper time is T.
- The time needed for the muon's travel is T×10.
- Time T×10 measured remotely at speed V becomes T.
- ***Therefore, in the muon's system the time T×10 becomes T.***
- Therefore, the muon can reach the Earth.

The assertion that attributes the duration measured remotely to the muon's proper time is absurd, unjustified and false as would be absurd the reasoning of a car driver who hopes to shorten a distance by measuring it remotely when he must travel it with all the necessary wheel spins. This reasoning is clamorously wrong, and is based on an overwhelming misunderstanding in the use of the notion of proper time. Given the theory, in the proper system of the muon the time of the muon's life is T, the time to reach the Earth is T×10, the time to travel the distance l if measured remotely becomes T (maybe), but the time it takes to travel distance l remains T×10.

There is no reason at all to authorize considering the time T×10 in the muon's proper system. Given the theory, if the travelling muon could measure the Earth with an instrument and developed calculations with the criteria of ballistic theory, it would calculate the distance l×0.1 and calculate the durations of Earth phenomena multiplied by 0.1. But given the theory, the proper time T of the muon does not change because it is in motion, and the distance l does not change when the muon, instead of measuring it remotely in a system other than its own, travels it metre-by-metre in its own system and puts itself in the condition, if it wants, to measure it with a ruler.

In other words: according to the theory, if the muon, considering itself at rest, measured remotely the time necessary for the Earth to come to it, it would find T. But if it deduced that its lifetime T is sufficient to reach the Earth, it would only show that it did not study carefully the theory of special relativity, because according to the theory the time required to reach the Earth is and remains T×10.

Above I simplified only the numbers and for the rest I reproduced the argument as it is read in a textbook used for university teaching, that of Rafael Ferraro. Now you will read the author's text. To understand the quotations we need to know that the half-life of a particle is the time it takes for half the particles of a quantity to decay into another element[66]:

> One of the first experiments that proved the phenomenon of time dilatation and length contraction (Rossi and Hall, 1941; Frisch and Smith, 1963) consisted in the observation of populations of high-energy cosmic-beam muons, to different altitudes above sea level. The muons are unstable particles that disintegrate to give an electron, a neutrino, and an antineutrino. The muon half-life is $T=1.53\times10^{-6}$ s. This means that an initial population of muons will be reduced to half after that time (in statistical terms). This period has been measured for low-energy muons, i.e., muons approximately at rest; so T is a proper time because

---

[66] Ferraro 2007, pp. 52-53.

it is measured in the proper frame of muons. Frisch and Smith examined populations of high-energy muons that are produced by cosmic beams entering the atmosphere (the muons' speeds were 99.52% of the speed of light). While the muons travel toward the Earth surface, they disintegrate; in this way the minor the altitude above sea level, the lower the population of muons. The experimental results indicated that the population diminishes $(27.5\pm3.0)\%$ for an altitude difference of 1908 m. Notice that a muon traveling at the speed of 0.9952 c covers 1908 m in a time Δt=1908 $m/(0.9952\ c)$=6.39×10^{-6} s, i.e., around four times its half-life. According to Newtonian notions of space and time, the population after that time would be reduced to $(1/2)^4$=1/16 of the original population, a reduction much greater than that really observed. However, since the mentioned half-life was measured in the proper frame of muon, we should transform Δt to the proper frame before making statements about the reduction of the population. In the proper frame of muon, the trip demands a time

$$\Delta\tau = \sqrt{1 - \frac{V^2}{c^2}}\,\Delta t = \sqrt{1 - 0.9952^2}\ 6.39\times10^{-6}\,\text{s}$$

$$= 6.25\times10^{-7}\,\text{s}$$

Then the population would be really reduced by a factor $(1/2)^{\Delta\tau/T}$=$(1/2)^{0.41}$=0.75; i.e., it will diminish 25%, in excellent agreement with the experimental result.

The experiment here examined is nothing but a realization of the situation described in Figure 3.1, where the particle (the muon) covers the proper length L_0=1908m in 6.39×10^{-6} s, at the speed V=0.9952 c. In the proper frame of the muon the length is contracted to L=$(1-0.9952^2)^{1/2}$ 1908m=186.72m, being then covered in a time of 6.25×10^{-7} s (at the same relative speed). In this way, time dilatation and length contraction are two complementary visions of a same physical fact.

In the quoted text there are two statements not perfectly coincident, or better, calibrated to give different suggestions: the first is that the muon's times dilate at high speed, and therefore its duration contracts and its phenomena slow down, and hence the muon has time to cross the 1908 metres of the atmosphere. We are fully in Lorentz's framework, where we can assume that it is the muon which contracts and not the Earth. This statement exploits the fact that given the smallness of the muon, the readers are always instinctively prepared to consider the muon in motion and the Earth at rest, although they know that speed is relative and that the choice of the frame at rest is arbitrary. The second statement, however, is that the apparent contraction of the length of 1908 metres is such to allow the muon to rejoice like the driver mentioned at the beginning of this section. The suggestion is to confuse the remote measure with the local one, because it cannot be suggested that the Earth really contracts with respect to the muon, since anybody would think that while the Earth moves to 0.9952 c with respect to the muon, it moves also at countless other speeds far smaller than that with respect to Jupiter, Mars, the Sun, and so on.

This was the account that we can read in a recent book. Given the tremendous importance of the case, I'll quote now the same pattern of reasoning that we find in a treatise of 1958. In this case we have to deal with the decay of the pion, another particle. Let us remember again that the half-life of a particle is the time it takes for half of the particles of a quantity to decay into another element. The reasoning here is as follows.

Premises:

- the half-life of pions determined by experiments at rest has the duration T;

- a set of pions can be accelerated to 0.99 c;

- given the half-life T, it is estimated that pions travelling at 0.99 c should halve their population after having travelled 5.3 metres;

- Instead, experiments show that the pions' population is halved at 39 metres distance from the source of a beam of pions.

How to explain this difference?

- the proper time of pions is T;

- but the apparent time measured by the laboratory is not T, but applying the gamma coefficient is T×7.35. This value T×7.35 is "non-proper". In fact, "moving clocks appear to run slow", as the author's text will teach us;

- but in time T×7.35, pions can travel exactly 39 metres, which explains the result of the experiment.

The contradiction here too is obvious: the pion in its system lives a time T, but if another system met it at a certain speed, the measure of that time would become T×7.35. So, it is said, the pion in its frame lives T×7.35: ***absurd and false conclusion***, and moreover asserted after expressly saying that the clocks in motion "appear" slower than at rest. In other words: a man lives a hundred years measured with his clock. But if we make a film of his life and watch the movie with a device that takes a thousand years to reproduce it, the show lasts for a thousand years. So the man lived a thousand years, measured with his clock that in the film slowed down by 90% like anything else: so a stupid conclusion that one feels uncomfortable while saying.

In this example, besides calculating the non-proper time and transforming it into proper time without any reason, but indeed with a flight of imagination worthy of the best science fiction, also the concept of reciprocity between time dilatation and durations contraction is misunderstood. This does not matter because to properly read the text it is sufficient to understand that the time was multiplied by 7.35, so the duration of the pions' frame was divided by 7.35, so the clock and the life of the pions slowdown of factor 7.35. However, this second misunderstanding has a function: to contribute to the disorientation of the readers, who must not become aware of the fundamental misunderstanding between proper and non-proper time. The secondary misunderstanding makes it even more difficult to verify the exact meaning of things that are said.

Here is the text. Note that the argument is presented to the reader in the form of an exercise, and therefore as an accessory of the main discourse, despite the major importance of the dilemma between apparent and realist interpretation. To understand the text, consider that the "target" is the bombarded element to generate the pions, and thus it is the origin and not the destination of the pions' movement[67]:

> Among the particles of high-energy physics are charged pions, particles of mass between that of the electron and the proton and of positive or negative charge. They can be produced by bombarding a suitable target in an accelerator with high-energy protons, the pions leaving the target with speeds close to that of light. It is found that pions are radioactive and, when they are brought to rest, their half-life is measured to be 1.77×10^{-8} sec. That is, half of the number present at any time have decayed 1.77×10^{-8} sec. later. A collimated pion beam, leaving the

---

[67] Resnick 1958, pp. 75-76.

accelerator at the speed of 0,99 c, is found to drop half its original density 39 m from the target.

a) Are these results consistent?

If we take the half-life to be 1.77×10^{-8} sec and the speed to be 2.97×10^{8} m/sec (= 0,99 c), the distance traveled over which half the pions in the beam should decay is

$$d=vt=2.97 \times 10^{-8}\ m/sec \times 1.77 \times 10^{-8}\ sec=5.3\ m.$$

This appear to contradict the direct measurement of 39 m.

b) Show how the time dilatation accounts for the measurements.

If the relativistic effects did not exist, then the half-life would be measured to be the same for pions at rest and pions in motion (an assumption we made in part a above). In relativity, however, the nonproper and proper half-lives are related by

$$\Delta t=\frac{\Delta\tau}{\sqrt{1-v^2/c^2}}.$$

The proper time in this case is 1.77×10^{-8} sec, the time interval measured by a clock attached to the pion, that is, at one place in the rest frame of the pion. In the laboratory frame, however, the pions are moving at high speeds and the time interval there (a nonproper one) will be larger (moving clocks appear to run slow). The nonproper half-life, measured by two different clocks in the laboratory frame, would then be

$$\Delta t = \frac{1.77 \times 10^{-8} sec}{\sqrt{1-(0.99)^2}} = 1.3 \times 10^{-7} sec$$

This is the half-life appropriate to the laboratory reference frame. Pions that live this long, traveling at a speed 0.99 c, would cover a distance

$$d=0.99\ c \times \Delta t=2.97 \times 10^{-8}\ m/sec \times 1.3 \times 10^{-7}\ sec=39\ m,$$

exactly as measured in the laboratory.

As far as experimental particle measures are concerned, we have nothing to say. Perhaps the experimental data is wrong, maybe it was poorly interpreted, or perhaps the particles in question have abnormal behaviours whose cause is unknown, or perhaps it is incorrect to identify the same particle in the two experimental contexts. About muons and pions we know nothing. We only want to note that the theory of special relativity cannot be used consistently to interpret any experimental data, and that authors attempting to use it introduce macroscopic logical errors in the chain of their deductions. Here a miserable rhetoric is used: since particles are elementary things, readers are led to ignore that even for a particle the distinction between space measured with a ruler and space resulting from remote measurements in motion by means of waves is valid according to relativity fundamentals. The meson has the same fate as the car driver of the first example: the space in which it moves is the proper space, the length of its life is inside its proper time, and the distance between itself and the Earth measured with a laser gun or something like this is not its proper space. The fact that a meson is an elementary entity and not an "observer" able to do measurements with a laser telemeter does not change the definition of the notions of proper space and time.

To express these arguments based on experimental particle data and to interpret them as proofs of relativity, it is necessary to create an ambiguous and misleading sense of the terms "proper space" and "proper time", it is necessary to suggest that things are meaningful and coherent when one states that space and time dilatations concern only remote measurements by means of electromagnetic signals, and in the same pages to state that dilatations have physical reality not depending upon the act of measuring. The whole is so immensely contradictory so as to be disconcerting and to make it difficult to believe that science can fall down to the point of producing similar arguments.

One last observation regarding μ mesons is that anyone who wanted could use the argument in favour of Lorentz's theory. In fact, given Lorentz's theory if a meson passes through the atmosphere at a measured speed v slightly smaller than that of light, it travels the ether at a speed w different from v (w is greater than v if the sense of the motion of the Earth is the same as the meson's, while it is less than v if the sense is opposite). On the Earth is measured v instead of w because the measuring instruments are in motion together with the Earth, but the order of magnitude of the contraction of meson durations for both v and w is sufficient to explain that it can reach the Earth. Let us admit that there is no mistake in mesons' observations, and that there is no way of interpreting their phenomena as interactions with the atmosphere (as it may be expected): in this case we would have no other explanation that the Lorentz's hypothesis. It is clear that neither this nor any other is the opinion of this book, whose purpose is simply to clearly describe the coherence problems of relativity.

6.3   *Minkowski geometry, space-time and world lines*

The representation of space-time through Minkowski geometry is a classic *pièce de résistance* of relativistic literature. We shall describe it now by giving the shortest description sufficient for our purpose, because anyone who wants to study the details about it will find no difficulty in understanding the descriptions that can be found in intermediate and upper level textbooks. To understand the Minkowski geometry, which is conceptually a simple matter, it is enough to know some analytic geometry from high school. If one reads it without thinking of the problem of its physical meaning, Minkowski's elegant geometry is a pleasant entertainment, and those born devoid of the gift or the burden of the sense of physical reality can find in it a beautiful description of the whole Universe.

Minkowski geometry is used to find a solution to the symmetry problem, or more precisely to find an objective reason for asymmetry and use it to decide to which to attribute space and time dilatation when there are two systems in relative motion. We shall then see that the conclusions drawn from Minkowski geometry are decisively and expressly on the realist side. It is argued, with the rigour that the readers will soon judge by themselves, that given two systems in motion one exists that undergoes time dilatation unlike the other. Readers will think: but this contradicts the cornerstones of the theory, and among other things it contradicts the intentions of the meson's argument, which tried to infer realist conclusions by saving the symmetry between two systems. This is so: the argument based on Minkowski geometry introduces asymmetry, stating that given two systems in motion one exists in which clocks are delayed because the systems are not equal, and frustrates the tens or hundreds of pages that preceded it in each treatise preaching the principle of relativity and the impossibility of distinguishing absolutely between rest and uniform rectilinear motion.

Minkowski geometry is articulated by defining a quantity called *space-time interval* and certain pseudo-geometric entities called *world-lines* or also *life-lines*. The term "pseudo-geometric" is neither ironic nor derogatory: it refers to the fact that these entities are described by analytic geometry equations but are impossible to represent literally in geometric intuitive space and time. The essential premise is that the statements of Minkowski geometry, together with its notion of the space-time interval and its notion of the world-lines, are a consequence following Lorentz's transformations that would apply to physical realities to which the application of Lorentz's transformations could be allowed, but are nothing more than this: they are a consequence, they do not convey any new information with respect to the premises, so they are no proof and cannot have any proof value regarding the reality of spatial or temporal dilatation. When we say the most well-known of the syllogisms, "All men are mortal, Socrates is a man, so Socrates is mortal", the consequence "Socrates is mortal" is necessary given the premise, but cannot be used to demonstrate the premise. Indeed, if a syllogism is based on incorrect premises, as it would be, "All men are rabbits, Socrates is a man, so Socrates is a rabbit", then the conclusion "Socrates is a rabbit" remains the necessary consequence of the premises but cannot be invoked to show that men are rabbits. To demonstrate this, we would need experience and arguments in which the proposition "all men are rabbits" (a proposition that if read in a psychological sense is very likely, but is not valid for Socrates) either is an empirical data point or is a necessary consequence of other true premises or assumes the character of a postulate, true by definition.

Likewise, Minkowski geometry, the notion of the space-time interval and the notion of world-lines are necessary consequences of Lorentz's mathematics, but cannot be used to demonstrate that Lorentz's mathematics have a physical meaning with respect to anything. There are some who argue that these consequences "proof" the theory of special relativity, and those arguing so urgently need the medicine represented by any elementary textbook of Aristotle's logic or modern mathematical logic, which in relation to this kind of problem give both the same teaching.

Hermann Minkowski, an immigrant from Lithuania, was a teacher at the Federal Polytechnic of Zurich and Einstein was a pupil of him, apparently indolent. When the 1905 papers began to circulate Minkowski was surprised, because in his judgment Einstein was a lazy fellow[68]; yet Minkowski became enthusiastic about relativity

---

[68] For historical information, See Isaacson 2007, Chapter 3.

and formulated a four-dimensional geometry model corresponding to a world in which the space and temporal coordinates are correlated through Lorentz's transformations.

We have already noted that Lorentz's transformations have a trend so that we lose in time what we gain in space with respect to the corresponding Galilean transformations, and vice versa. For this reason, relativity does not speak any more of space and time, but of space-time: the three spatial dimensions and the temporal dimension are no longer considered independent as they have always been, but are considered to be co-ordinated in a whole of four dimensions. Given the coordinates x,y,z and t, applying Lorentz's transformations we always have:

$$x' > x, \, y' > y, \, z' > z,$$
$$\text{but } t' < t.$$

Let us look again at the table with the numeric cases of the transformed values for speed v between 1 and 1,000,000 metres per second:

Speed m/s	x' Galilean	t' Galilean
	x' Lorentz	**t' Lorentz**
1	1,000,000,000.000000	100
	1,000,000,000.000000	**100**
10	999,999,100.000000	100
	999,999,100.000001	**100**
100	999,990,100.000000	100
	999,990,100,000056	**99.999999**
1,000	999,900,100.000000	100
	999,900,100.005555	**99.999989**
10,000	999,000,100.000000	100
	999,000,100.555000	**99.999889**
100,000	990,000,100.000000	100
	990,000,155.000010	**99.998894**
1,000,000	900,000,100.000000	100
	900,005,100.042223	**99.989444**

As we see, at a speed of one million metres per second, on a length of 900 million metres Lorentz x' is about 5,000 metres more than Galilean x', while Lorentz t' loses about a hundredths of a second in a hundred seconds. Given this property, it can be shown that this relationship is valid:

$$x^2 + y^2 + z^2 - c^2t^2 = x'^2 + y'^2 + z'^2 - c^2t'^2;$$

and therefore can be defined a quantity:

$$d^2 = (x - x')^2 + (y - y')^2 + (z - z')^2 - c^2(t - t')^2.$$

This quantity is called the *space-time interval*, which is invariant: the same events represented in systems K, K', K" and so on have the same interval. Interval has an analogy with the distance between two points in the Euclidean space. In fact, let us remember that in the two-dimensional Euclidean space the distance is:

$$d = \sqrt{(x - x')^2 + (y - y')^2}$$

and in three-dimensional space it is:

$$d = \sqrt{(x - x')^2 + (y - y')^2 + (z - z')^2}$$

Here we have a pseudo-distance:

$$d = \sqrt{(x - x')^2 + (y - y')^2 + (z - z')^2 - c^2(t - t')^2}.$$

To easily understand the concept of space-time interval and Minkowski geometry, we must go back to the conventional or apparent interpretation of relativity and think that the x,y,z,t coordinates of a system K and x',y',z',t' of a system K' are the coordinates of the same event evaluated by systems in motion with respect to each other. As we know, according to relativity time t is different than t' because the K' system makes measurements based on the reception of light signals from the K system and the result is deformed by the constancy of c, and the spatial coordinates are different between the two systems because the clock synchronization in the two systems affects the space measurements as well as those of time.

Now, the space-time interval is the square root of an amount that can be less than zero if

$$(x - x')^2 + (y - y')^2 + (z - z')^2 < c^2(t - t')^2.$$

Thus the space-time interval can be real or imaginary and the meaning of the two cases is this: if the interval is real, the values of t and t' in the two systems are such that the event can be known in both systems at the time respectively t and t', because light had time enough to bring the information about the event from one system to another. If the interval is imaginary, an event occurring in K at time t cannot be known in K' at any time t', because the signal attesting the event cannot yet have reached K'. The definition of space-time interval also implies another consequence: the simultaneous events of which moving systems become aware through the exchange of electromagnetic signals have different time coordinates in the different systems, but are always considered in the same temporal order by all the systems in motion where the interval is a real number. If event A takes place in K at time t_A before event B occurring at time t_B, and so in K we have $t_A < t_B$, then also in K' we have $t'_A < t'_B$. The difference between the two time values changes in the two systems as a result of dilatation, but the order of events is always preserved. Incidentally, it is always observed that for this reason special relativity does not call into question the classical notion of causality, but for us this has no significance either good or bad, given the radical logical inconsistency of the theory.

The space-time interval introduces us to the idea that in place of the classical representation of the three-dimensional space geometry next to the one dimensional temporal geometry, we can represent the four dimensions in a single geometry, which is called geometry of space-time. But this will be a non-Euclidean geometry not only because it has four dimensions, but also because in it the straight line is no longer the shortest connecting line between two points, but it is the longest line connecting two points. This follows from the form of the space-time interval definition. It is shown that the calculated length of each curved line connecting two points is shorter than the length of the straight line connecting them, and therefore Minkowski geometry consists only of equations of analytic geometry and cannot be represented in the Euclidean space with intuitive examples that literally correspond to the properties of the represented figures: intuitive graphical representations are only useful analogies that will help to understand this purely intellectual geometry. This happens both because in the Euclidean space we cannot represent four dimensions, and because in the Euclidean space we cannot represent the fact that the straight line between two points is the maximum distance between two points rather than the minimum.

Now we shall build step-by-step an example of Minkowski geometry: at each step we must remember that the straight line joining two points is longer than a polygonal or curved line that connects them. We must remember to always think that the straight line is longer, although in the graphic representation we perceive it shorter, and therefore graphic representation should not be taken literally, but only as an auxiliary tool for understanding. To simplify the example, we only construct the representation for x and t dimensions, because if one understands the key of this geometry, then imagining the extension to the case of three or four-dimensional space-time follows immediately.

To construct a two-dimensional space-time geometry, we must set a Cartesian axis system in which the units of measurement are chosen so that the path of the light be inclined by 45°, so we can, for example, set as time unit the second and as space unit a length of 300 million metres:

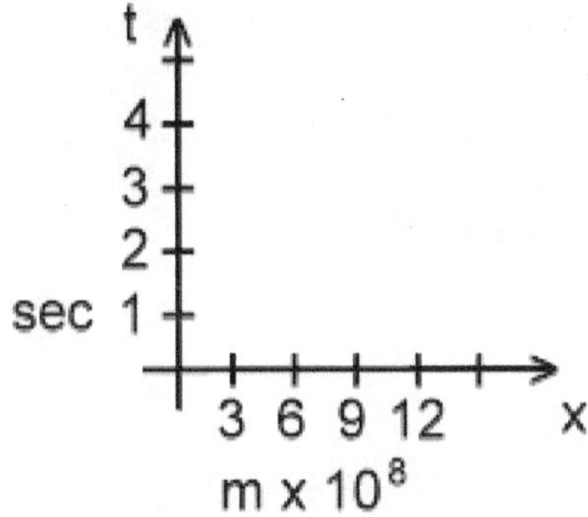

On the x-axis we have units equal to the distance covered by the light in a second, on the t-axis we have the time in seconds. Obviously, in this geometry, the path of electromagnetic signals is represented by 45° sloping straight lines, and the straight line inclined 45° and passing through the origin is called the light line:

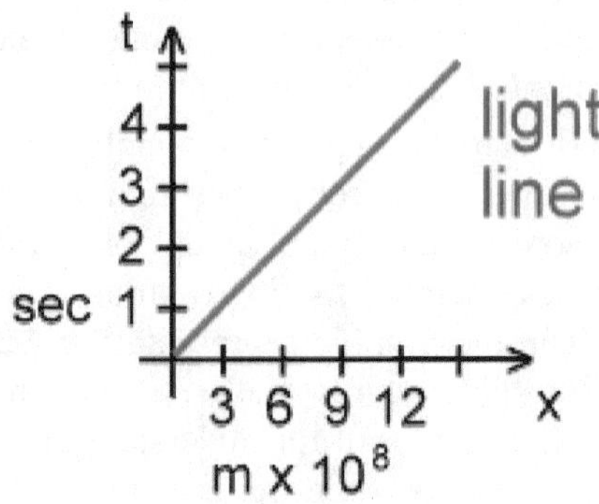

In this geometry we can represent lines, called *world-lines*, that connect the events of a system considered at rest or in motion. The representation has typical features according to the cases. For example, let's consider the following two events: a man writes on a sheet the letter A and shortly afterwards he writes letter C. These events can occur while the man is considered at rest, or while the man is considered in straight-line uniform motion in an aircraft flying at constant speed, or while the man undergoes a strong acceleration because he is in an aircraft that is taking off.

If events are represented at rest, the man's world-line between the two events is a straight line parallel to the axis of time, because event C is next to A — there is time between them — but there is no space gain because the man is not in motion:

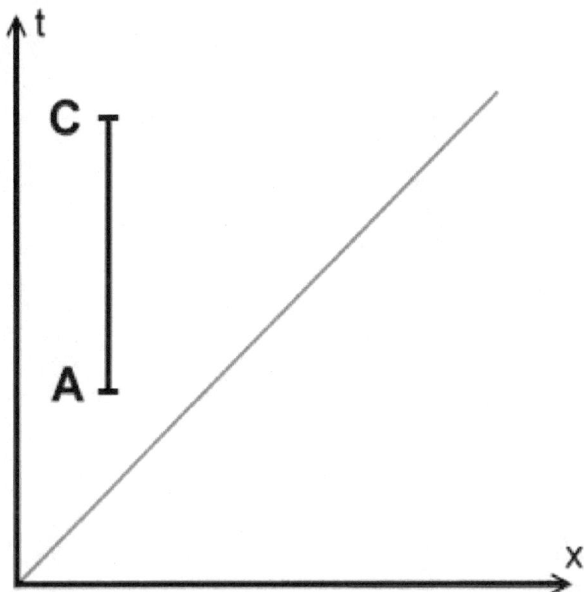

If the events are considered in motion at constant speed, the world-line is still a straight line but inclined. In fact, in this case as we age in time while we move in space and the coordinate x changes as a function of t. The straight line can never have an inclination equal to or smaller than 45°, because the speed of light cannot be reached. So it is not possible for a world-line to be parallel to the light line. This is the figure for rectilinear uniform motion:

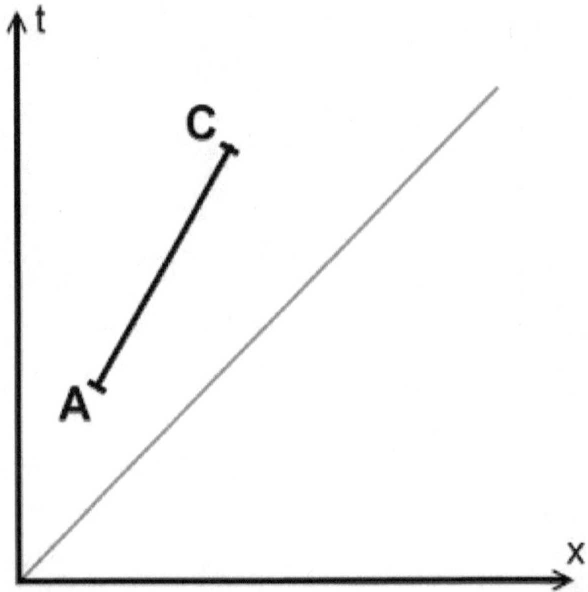

Finally, if the motion is accelerated, the world-line will become curved, because as time increases we gain speed, and then the space travelled per time unit will increase:

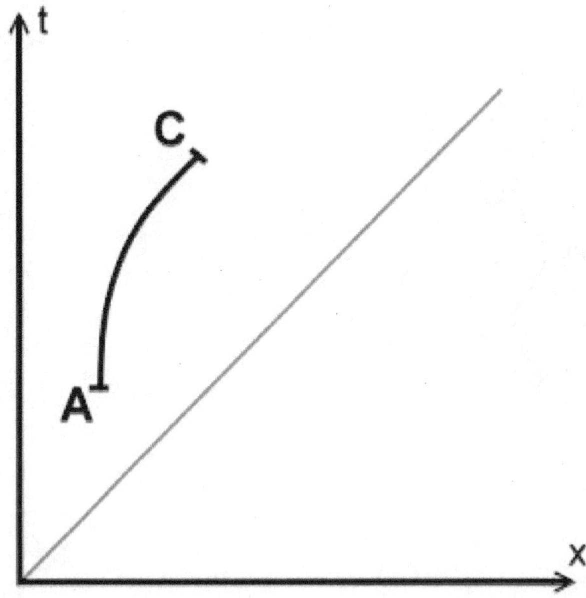

In this example, between the two events A and C the system has reached a considerable speed, but anyway the line cannot reach the inclination of the light line, i.e. the speed of the light.

Finally, consider the world-line of the day of a tram moving from terminal A to terminal C several times between morning and evening. If the path were perfectly rectilinear and therefore the graph could be drawn with only one spatial dimension, the representation would be this:

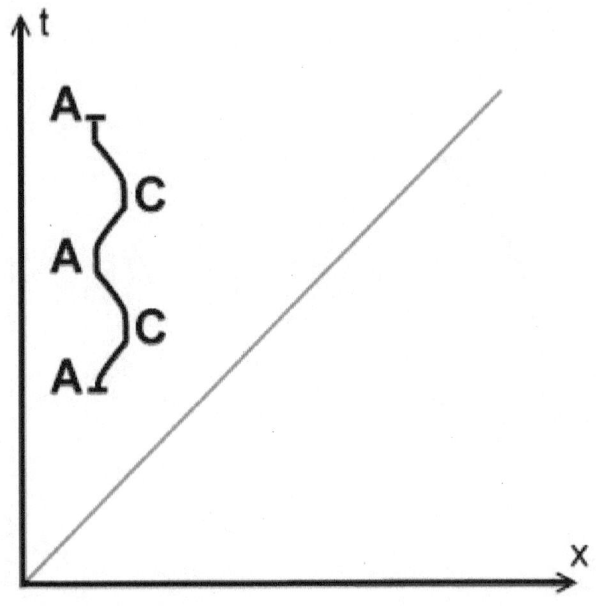

The tram leaves terminal A, accelerates and then proceeds at constant speed to C, stands in C for a while, then returns to A, and so on.

The most important thing, which corresponds to a certain mathematical formalism, but that is sufficient to comprehend in a generic way without resorting to that formalism, is the following. In this geometry, a function is defined to calculate the distance between two events, that is the length of a straight or curved line. The distance of two points connected by a straight line is calculated using the space-time interval, while the distance along a curved line is calculated by solving an integral of the intervals of the infinitesimal segments that make up the curve[69]. As we have already said, in this geometry the straight line represents the maximum distance between two points, and therefore the function calculating the distance returns a lower value for curved lines than straight lines, and returns the maximum value for straight lines. For example, consider two successive events, A and C, which consist of the fact that a man F

---

[69] To deepen quickly this issue, see Fabri 2005, pp. 104-108.

writes letters A and C one after the other on a sheet and with a certain time between them. Man F is travelling on a train at constant speed, so his world-line is straight. Before F writes the letter A, next to F is sitting another man we call M. Just as event A happens, the man M rushes along the train at a certain speed, then he reverses the sense of motion and returns to F exactly when F writes the letter C. At time of event A, the x and t coordinates are the same for F and M, and the same happens at the time of event C. But in the time between the two events the world-line of M is curved because he accelerated and then decelerated, while the F line as we know is straight given the constant speed.

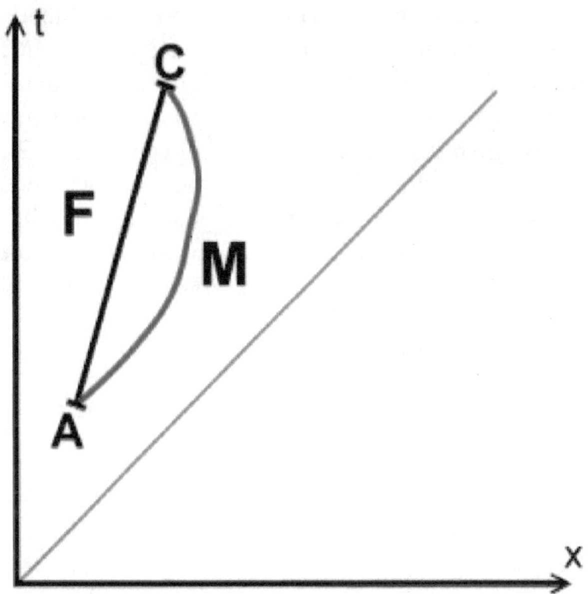

The Euclidean and intuitive geometric representation of the figures does not literally correspond to the equations of space-time geometry, because in figures the length of line F appears to perception necessarily smaller than that of line M, while in space-time geometry the function calculating the length of the straight line F gives a greater result than all the non-straight, curved or polygonal lines. This feature is a representation of the reciprocity between spatial contraction and temporal dilatation resulting from Lorentz's transformations: the man M travels more than the man F in space-time, so the space travelled by him is contracted (the space containing his journey is contracted, not the space of his dimensions, which is dilated as time) and his time is dilated and his clocks and all phenomena of his system proceed more slowly than those of the F system.

The ultimate feature of space-time geometry is that it allows us to represent simultaneous events easily. When F and M are in the same place, they detect events A and C at the same time directly. But when M moves away from F, the intermediate events of F between A and C are known by M when M receives a signal that witnesses them. Since the signal travels at the speed of light, events in F are attributed by M to temporal coordinates differing from those attributed by F. To find the coordinates of any event observed by F and M, it is enough to just plot the parallel line to the light line. Suppose that between the writing of letters A and C the man F puts the pen on his desk (event e1) and then takes again the pen in hand (event e2). If M observes F thanks to the light reflected by F, these events are known by M at the times $e1_M$ and $e2_M$, when their image reaches M at the speed c. As we know, time coordinates vary according to Lorentz's transformations, but the order of succession between events is always preserved:

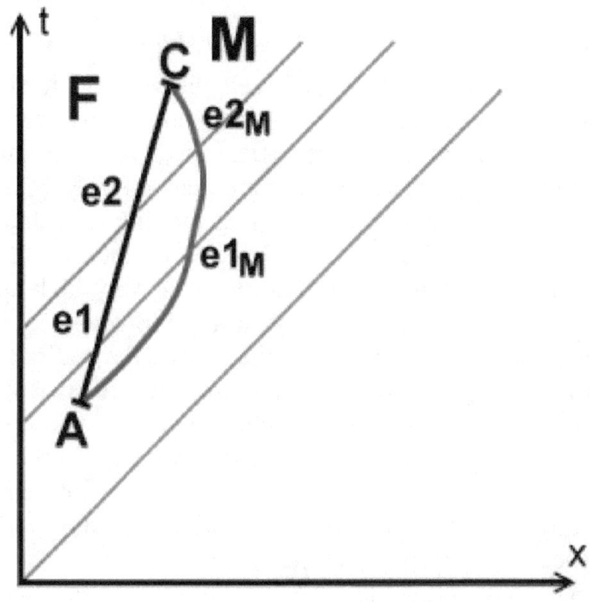

From the purely formal mathematical standpoint, this elegant representation is impeccable. If there are systems related through Lorentz's transformations, then this representation of their relationships follows from the definitions. We shall now see how this representation is used by relativistic literature to answer the question: given two systems in relative motion, in which of them is time dilated and clocks delayed in a real way, so that the difference in their indications can be verified with local measures when the systems rejoin?

6.4   Minkowski geometry and clocks problem

The μ meson argument which we analysed above serves to find an empirical confirmation of relativity while maintaining the symmetry between the systems in motion. The logical error is to attribute to proper space and time the measures that are guessed to be given by remote measurement, and the logical error is implemented and expressed by the rhetorical contrivance implied by the fact that the meson is a particle of minimal size that moves at very high speeds compared to the Earth. If the same way of arguing were developed with the usual examples of trains or similar things, the reader would point out that remote space and time measurements are attributed to proper space and time with a Pindaric flight that occurs tacitly during the argument. The same Pindaric flight can be occulted in the mists of the equivocation because the fact that mesons are particles provokes the irrational but powerful suggestion that the world of relativity is especially appropriate to them.

Minkowski geometry instead is used to justify asymmetry between two systems. Even before examining the argument, we can notice that contradiction already pervades the starting point, because after having asserted the symmetry between systems, now authors want to find a reason to justify asymmetry. Within the same books, authors use the μ meson argument to try to prove that the theory gives rise to physical consequences despite absolute symmetry, and then in another chapter they use the argument of Minkowski geometry to try to prove that symmetry is not absolute. Therefore, the reader will ask, the time dilatation of two systems A and B in relative motion is symmetrically the same for B measured by A and for A measured by B, or is not the same and there is a way to show that dilatation occurs just in only one of the two systems? No one is able to answer this question, which is one of the consequences of the general ambiguity and oscillation of authors between the apparent interpretation of the theory and the realist interpretation.

So let us just leave this question unanswered, and let us consider the solution to the clock problem that is developed on the basis of Minkowski geometry. To understand this argument, we must use the notion of privileged system. It has always been known, and has no relation to relativity, that if we have the need to solve certain problems, it is essential to fix a particular reference system for speed determination. That is, it may be necessary to consider speed only in a certain frame, without that this is a violation of the law of inertia. This so stated sounds strange, but a simple example clarifies it immediately. A car travels at a speed v on a road, and we know that given the law of inertia we have to admit the possibility of considering the Earth and road in motion and the car at rest, or the car in motion at speed $2v$ compared to a vehicle that crosses it in the opposite sense with the same speed, and so on. However, there are two main factors hindering the car's motion: aerodynamic resistance and rolling resistance of rubber wheels. The car has to spend energy to move the air in front and other energy for the spinning of its elastic wheels, which are constantly deformed by the weight of the car. Incidentally, rolling resistance is the reason why racing bicycles mount the most thin and rigid tubular tires possible and is one of the reasons why trains employ less energy than other vehicles thanks to the rigidity of their steel wheels. Air resistance and rolling resistance both increase with speed and impede motion with increasing force with speed, as it is intuitive. But what is the speed which matters? It is obvious that if we want to know what power is needed to keep a car in motion at speed v, we must calculate the rolling resistance according to the speed v with respect to the road, and the air resistance as a function of the speed w not with respect to road, but to the air: in fact, wind must also be taken into account, which can increase or decrease the air resistance. In this case, there are two privileged systems, the road and the air: for physical reasons not at all fictive, it would not make sense to calculate the resistance to motion by considering different systems, respectively, from the road and the air.

With this premise, let us consider two systems F and M. F is at rest or in rectilinear uniform motion, while M moves away from F along a straight line and arrives at some point x where it undergoes an acceleration to reverse the sense and returns to F in O:

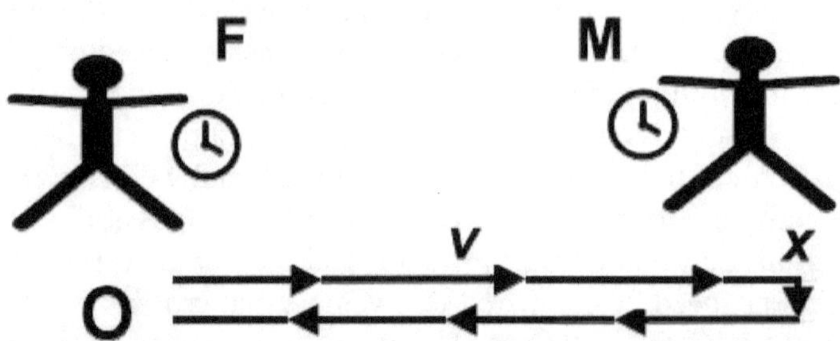

There is an objective element that distinguishes M from F: M undergoes at least one acceleration, while F is subject to none. The rectilinear motion that the two perform at constant speed is relative according to the law of inertia, but the acceleration element distinguishes the system M from system F in an objective way and with regard to acceleration the difference between the two systems is absolute. With this premise, that no one contests (or at least classical mechanics does not contest), we must go back to the relativistic literature, because without it we would not know what conclusions to draw.

Literature states that time dilatation concerns the M system which has accelerated at least once. We know that this argument had been mentioned by Langevin in 1911 by introducing the example of the two twins, and this argument now comes back. The first thing that comes to the minds of readers who try to find a logical consequentiality in the theory of special relativity will be: but then if time dilatation is a consequence of the acceleration, it will have to be determined as a function of acceleration according to formulas nobody has talked about, at least until now. But the relativistic literature knows that in this case Lorentz's mathematics would no longer be applicable, so it would be necessary to reformulate the theory with all other criteria and reopen the problem of Maxwell's equation and of the constant c. So the theory must find the way to argue that acceleration is the cause of dilatation, but dilatation is a function of speed. We need to build something that allows us to express this assertion, which contradicts the whole history of the scientific use of mathematics since Galileo's time, but which is vital to preserve the theory. The readers are surely aware of the enormity of the assertion that attributes the cause of a phenomenon to a certain magnitude and then writes the corresponding equations as a function of another magnitude: but this is exactly what we now encounter in relativistic textbooks.

Resnick proceeds this way[70]. He first warns that we should not consider the amount of acceleration[71]:

> First, we can simplify matters by ignoring the effect of the acceleration on the traveling clock. (...) The error thereby introduced [by acceleration] can be made very small compared to the total time of the trip, for we can make the trip as far and as long as we wish without changing the acceleration intervals. It is the total time that is at issue here on any case.

Then he develops a numerical example in which M travels at a speed close to c, and therefore applying Lorentz for system F will elapse 10 years, while in the system M will elapse only 6. In the example, both systems send to the other a light signal every New Year's Eve, so during the trip F sends 9 signals and M sends 5 because in the day that for F is the tenth New Year's Eve and for M the sixth, the two systems rejoin and send one to each other a last signal as both are again at rest in the same place. Represented according to the space-time geometry, the situation looks like the following figure; let's remember that the M system moves to-and-fro on the x-axis and that the representation takes this triangular aspect because we have time in the ordinate and F is at rest:

---

[70] Resnick, pp. 203-208.

[71] Resnick, p. 204.

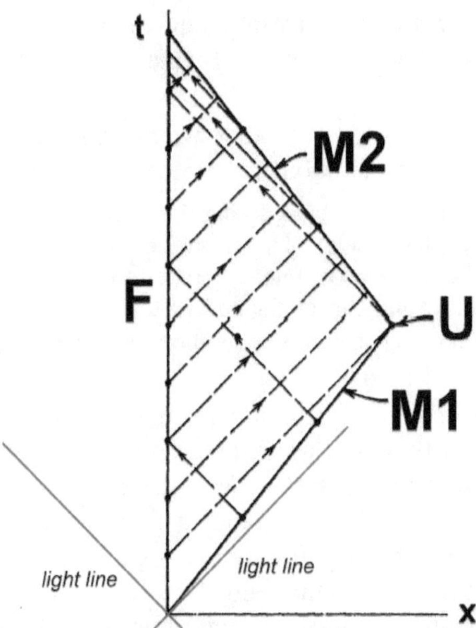

F is the world-line of system F and coincides with the axis of time, M1 is the world-line of M while travelling forth, U is the inversion point where M receives a negligible acceleration and M2 is the world-line of M on the return trip. The arrows represent the New Year's Eve signals from each, and are parallel to the light lines, because the signals travel at the speed of the light. Signals sent at New Year's Eve by each sender reach the recipient when they can reach it given the finite speed of light.

At this point Resnick's syllogism is:

- Premise (1): "Due to the time dilatation, the signals of each of the two reach the recipient with a predictable shift."

- Premise (2): "But the accounts will balance in every case: F sends 9 New Year signals and M receives them when they must reach it travelling at speed c, M sends 5 and F receives them when they have to reach it. So F counts 6 years of M, and M counts 10 years of F."

- Conclusion: "So the phase shift of the signals demonstrates the reality of the temporal dilatation of the M system."

We are again fully in a *petitio principii*: the premise (1) is invoked to deduce the conclusion, and the conclusion is invoked to deduce the premise (1). The premise (2) is irrelevant to anything, and it only serves to enrich the speech, which in the text is developed for several pages. Given what is stated in the premise (2), which again mentions the fact that each system "counts" in a certain way the years of the other, the astonished and perhaps perplexed reader might return to think that the theory describes a conventional fact, which would only happen if we synchronized the clocks by means of light signals with some special procedure. But now the text objects once more that this is not so, that temporal dilatation is a natural phenomenon: Resnick confirms the realist interpretation by explaining that M's biological clocks will behave like mechanical clocks and everything else, and that at the end M will be 6 years older than at the beginning of the journey while F will be 10 years older. So on the whole the argument can also be reduced to these assertions:

- given the mathematical form of Lorentz's transformations, follows the mathematical form of Minkowski geometry;

- so if a Lorentz transformation is applied to a physical system, then the properties of Minkowski geometry would apply to that system;

- given the form of Minkowski geometry, the path in space-time of the moving system M receives a different representation than the path of the system F at rest;

- Lorentz's transformations therefore have physical meaning and the realist interpretation for the M system is allowed.

Quare opium facit dormire? Quia est in eo virtus dormitiva![72] The arbitrary character of the argumentative technique is immense, yet all of these are things written by a man who was an academic of rank and breeding: consult an Encyclopaedia at the entry "Robert Resnick" to find out.

Ferraro uses a different technique. In the introductory pages of his book he defines the concept of privileged system in terms identical to

---

[72] Joke by Moliere in *Le malade imaginaire* quoted until being abused, but here unavoidable.

those used by us in this section, using the example of the resistance to motion exerted by a fluid, and takes care to specify[73]:

> Thus, the fluid becomes a privileged reference system. It is clear that this privilege cannot be taken as a violation of the Principle of relativity because the type of phenomenon under consideration *naturally* privileges the fluid frame.

In fact, what we do when a privileged system is assumed is not to declare motion absolute with respect to a given system, but it is to choose a given system as frame because the calculated quantities that concern a problem to be resolved are those relative to the interaction with that frame and not another.

When it comes to talking about the twin paradox, or clocks in relative motion, after developing the mathematical properties of space-time geometry, Ferraro states[74]:

> The question examined in the previous paragraph is usually introduced as "the twin paradox". One of the twins is astronaut and undertakes a space journey. At the departure the twins check their clocks. When the astronaut comes back the twins compare their clocks again and realize that the astronaut's clock went slower than that of his brother on Earth. Not only the clocks exhibit the discrepancy, but the astronaut came back younger than his brother. The reason for describing this result as "paradoxical" lies in the fact that, from the astronaut's point of view, he would have the right to consider that the traveler is not him but his brother; if so, his conclusion would be that his brother should be younger when they meet again. Thus we arrive to two contradictory conclusions, depending on whether the problem is analyzed from the Earth or the astronaut frames.

This is remarkable because at least it attests that even for the relativistic literature it is contradictory and unacceptable to say that clock A is delayed with respect to clock B *and* clock B is delayed with respect to clock A. Then Ferraro draws the figure representing the world-lines of F and M in the usual way, but without introducing the yearly light signals, which are not part of his demonstration technique:

---

[73] Ferraro 2007, p. 11.

[74] Ferraro 2007, p. 104.

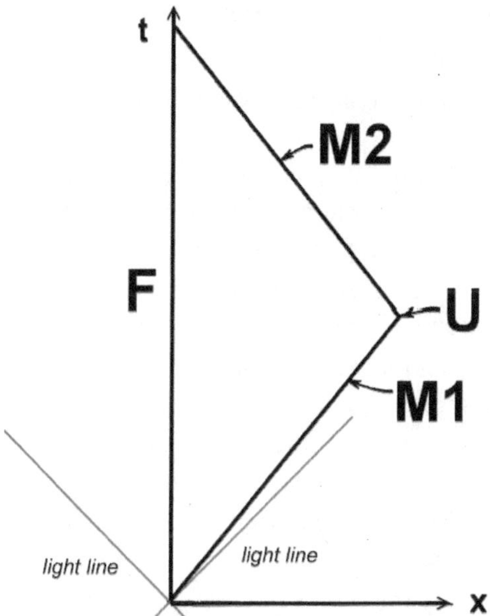

And on the basis of this representation Ferraro draws the following conclusion[75]:

This geometric framework confers a privilege to the straight world line (i.e., to the rectilinear uniform movement). In the same way that Euclidean geometry states that the distance is minimum along a straight line, the pseudo-Euclidean geometry of Special Relativity states that the time elapsed between two events, as read by a unique clock, is maximum along a straight world line. So, there is no symmetry in the behavior of the astronaut and his brother clocks because their relative movement must be non-uniform in order that the clocks can go away and meet again: if one of the clocks has a rectilinear uniform movement (straight world line), *necessarily the other will not have it*. Like in classical physics, in Special Relativity there are also privileged world lines: the straight timelike world lines corresponding to inertial trajectories of particles. The privilege of these trajectories is conferred by an underlying geometrical structure existing independently of the physical phenomena, and playing the same role that Euclidean absolute space and absolute time play in classical physics. The privilege of inertial world lines on the others explains the asymmetry between the

---

[75] Ferraro 2007, p. 105.

rates of the clocks carried by the astronaut and his brother.

So Ferraro's argument is:

- "To identify the system at rest, we would need a privileged system."
- "A privileged system cannot be conventional, but must correspond to the natural character of a phenomenon" (this was said by Ferraro, though 94 pages earlier).
- "But we define a representation method through a particular geometric projection technique that defines world-lines."
- "In this representation the distance between two events represented by a straight world-line is maximum."
- "But this representation exists independently of physical phenomena."
- "So anything represented by straight world-lines is inside a privileged system."
- "Therefore will delay the clocks of system M, whose world-line is polygonal."

I do not know if it is necessary to comment: the vague intention is to prove that the system without acceleration is that where time is not dilated, but Minkowski geometry *does not* define any privileged system, because a privileged system is given by the experimental study of *natural and experimental* features, as Ferraro says, of a phenomenon, not by a conventional mathematical definition that is incapable to create any information that is not already implied in the premises. Minkowski geometry perhaps is "existing independently of the physical phenomena", but it exists in the same way as if there are two apples the number 2 exists. If there are two apples, the number 2 surely exists, but also the number 4 corresponding to half apples exists, and the number 8 and so on. But given this, we cannot deduce anything about the physical reality of the two apples and about what will happen to them, because we did not add any information to what we knew: to conclude from Ferraro's reasons that the clocks of the M system will delay is as consistent as the inference of one who said that since there are two apples and there is number two corresponding to the two apples and there is number four corresponding to four half apples, then the two apples will be split into four halves without necessity of cutting them with a knife. And why not in eight or sixteen parts? And more, in Minkowski pseudo-geometric representation, the straight line is the longest line between two points, but in the Euclidean representation with which we depict the situation, the straight line is the shortest. Now, even the Euclidean representation is a "geometrical structure existing independently of the physical phenomena"; we do not see why classical representation has less rights than that of Minkowski to this qualification. So why do not we conclude that the F system is the one whose clocks are delayed, instead of M? After all, we would have the same bad reasons to draw this conclusion instead of the other. Given the Lorentz transformations, it would be enough to define with a little fancy an anti-Minkowski geometry that mathematically comes out from transformations, but has an opposite form, to draw the conclusion that the delaying clock is the one that does not accelerate rather than the one which undergoes accelerations. In all this, there is no trace either of study of the original problem or of curiosity about nature: there is no interest at all to progress a step forward in the knowledge of propagation of electromagnetic waves or any other natural fact, but the goal is to continue to assert an attempted and failed solution because for extra-scientific reasons it has been raised to a symbol.

Ferraro's argument adds nothing to Langevin's first assertion in 1911 (section 5.5): in both cases acceleration is invoked to exploit the fact that classical physics associates it with the objective element of force. The difference is that Langevin mentioned acceleration while talking to a distracted audience without adding explanations, while here the use of acceleration is justified by having invented an ingenious and unusual way of representing accelerated motion: Minkowski geometry gives no new information, and invoking it as an evidence has the same argumentative force as that of someone who could use the (true) argument that the word "acceleration" has 12 letters. However, if we really wanted to make the effort to accept Ferraro's deduction, the known questions would remain unanswered. First of all, the systems in the play are always two in isolated examples depicted by the theory, but are infinite in the world. Let us imagine this situation: F is at rest, M moves to the right as in the above example at speed v, but there is a third system P that moves at speed *w* lower than *v* to the left, and passes through point F when also M is close to F. P does not accelerate because it moves infinitely on a straight line at speed *w*. M moves away from F at speed *v* and arrives at point U. But during the travel it also moves at speed *v+w* compared to P. Which speed is to be considered to calculate the time dilatation of M? Should we compute it for *v* or *v+w*? Then M inverts the sense of motion and moves toward F at speed *v*; the speed sign has changed, and the speed is now -*v*, but suppose now we overlook this problem. M arrives at F and finds that the clock has delayed. Then, since *v* is greater than *w*, M continues to travel to P at speed *v-w*, and reaches it. Is the contraction detected again? And with respect to which speed should it be calculated?

A remarkable fact is that the length of the arguments can distract the reader from the fact though obvious that the systems in motion in the world are not two as in the examples of textbooks, but are infinite. If we also choose to overlook this problem, there remains another, whose statement forces unintentionally comic things. According to the authors, the dilatation is due to acceleration, but depends on speed: in fact, everyone knows that if dilatation occurred as a function of acceleration, it would be necessary to rewrite the theory from the beginning for Maxwell's equation problem. Let us remember now that a single acceleration is needed to create the twins' situation, i.e. the one that reverses the motion, because on both arrival and departure the two characters could be in motion relative to each other and their coordinates could only coincide in two initial and final instants in which the clocks are compared locally in a minimum time. Then, let us ask, when the system M is in motion before reaching the point U of inversion of the sense, the dilatation happens progressively with continuity, following the Lorentz transformation applicable to the case, or not? If dilatation progresses gradually, we must make the extra assumption that all things, including a stone or an electron or a monkey's body, have the quality of prophecy in relation to the future accelerations they will undergo: they dilate because it is written which accelerations everyone and everything will undergo, and perhaps all physical systems know what is written. So[76]:

---

[76] Matthew 4, 6-7.

γέγραπται γὰρ ὅτι Τοῖς ἀγγέλοις αὐτοῦ ἐντελεῖται περὶ σοῦ καὶ ἐπὶ χειρῶν ἀροῦσίν σε, μήποτε προσκόψῃς πρὸς λίθον τὸν πόδα σου.

for it is written, He shall give his angels charge concerning thee: and in their hands they shall bear thee up, lest at any time thou dash thy foot against a stone.

The appropriate answer is only one:

ἔφη αὐτῷ ὁ Ἰησοῦς, Πάλιν γέγραπται, Οὐκ ἐκπειράσεις κύριον τὸν θεόν σου.

Jesus said unto him, It is written again, Thou shalt not tempt the Lord thy God.

If dilatation does not happen progressively, but suddenly occurs at point U, we must make a different extra assumption, that is, that the stone, the electron and the monkey body have an infallible memory, and remember the speed with which they have travelled before undergoing acceleration. How else could dilatation be a function of speed? We cannot escape these considerations: if a phenomenon depends on a magnitude A that is its cause but varies as a function of a magnitude B that is not A, then the phenomenon must have something similar to memory and intelligence to adapt the variations of B to those of A. Otherwise, physical laws will be written according to the magnitudes that are considered to be the cause of the phenomena, as has always been done.

This applies to the simplified example where there is only one acceleration at the moment of motion reversal, as well for any real case where there are innumerable accelerations: in order that the contraction of durations and lengths can be caused by acceleration but calculable as a function of speed, the system subject to dilatation must have either memory or prophecy or both.

A radical relativist might still say: dilatation occurs instead when the system M rejoins F, because then it is "observed". So the stone, the electron, and the monkey body in this case should have the ability to know that they are observed by some relativistic scientists who measure their accelerations, and remember to contract when rejoining with their twin at rest.

6.5   *Fizeau's experiment with water*

Fizeau's experiment, which disassembled a light beam and made it pass through pipes where water was moving at a controlled speed, gives us the opportunity to detect a set of implicit assumptions and superficial assertions that are consolidated in the theory. Fizeau's experiment is invoked by Einstein in the popular account of relativity[77] and occurs often in literature after Einstein. Pais[78] reports that the experiment is part of a set of experimental data that Einstein had taken into account before 1905. Here we describe the experiment by cutting it to bone because this is useful in order not to miss the basic logical architecture of the argument, but those who want to deepen their understanding can consult Einstein's quoted source and, for example, Ferraro's discussion[79].

So far, to avoid complicating our discourse with not indispensable data, we did not talk about two issues which at this point, however, it is indispensable to consider.

First, it is experimentally established and undisputed notion, independent of relativity, that c is the speed of electromagnetic waves in vacuum (or ether, according to the ancient interpretation), and that when electromagnetic waves cross a medium they move at a lower speed than c, determined precisely by the physical characteristics of the material medium that is crossed and interacts with the waves. Waves move at a speed slightly less than c in the Earth's atmosphere, while they move at a speed of about 2/3 c in water. Specifically, the speed of light in a fluid is a function of the refractive index, that is, the angle for which a beam is deflected when entering into air or water.

Second, in Lorentz's theory, an obvious consequence of the contraction of distances and lengths is that nothing can reach the speed of light, because the contraction intervenes to prevent it: the Lorentz factor's trend represents this fact. This gives rise to the consequence that we cannot compose two speeds simply by adding

---

[77] Einstein 1916, Chapters 13 and 14.

[78] Pais 1982, Chapter 6 and elsewhere.

[79] Ferraro 2007 treats the issue in sections 2.7 and 3.6.

them, as we have already seen above in section 2.8. If a person moves at 5 m/s in a train that moves at 100 m/s, we shall say that compared to the Earth, the person moves at a speed of 95 m/s or 105 m/s according to the sense of its motion within the train. But in the Lorentz framework this is an approximation: given two speeds v and w, to find the overall speed u we cannot apply the simple sum of speeds and say $u=v+w$, because the bodies moving at v and w speed are both subject to contraction. Therefore, the formula for composing two speeds is not the simple sum, but it is:

$$u = \frac{v + w}{1 + \frac{vw}{c^2}}.$$

Let us look at what follows from this formula in the Lorentz's theory:

- If two bodies moved the first at speed $v=c$ and the second at speed $w=c$ with respect to the first, the composite speed would still be c: just make the substitutions for immediate verification. This case, however, does not represent any real situation.

- If the two bodies moved at two speeds near c, the composite speed would also be slightly lower than c, and would therefore be much smaller than the sum of $v+w$, which would give a result almost equal to $2c$. In the framework of Lorentz this does not represent a paradox: if the first system moved at a speed v smaller than c in the ether, then the second system could move at a speed w slightly lower than c when measured with respect to the first system, but it would be subject to a very strong contraction of its lengths and duration, so the speed of the second system measured with respect to the ether would be much lower than w. In the second system, the celebrated twin would live actually 2 years, while in the first his brother would live 200.

- If the two bodies moved at terrestrial speeds, as is the case with the man on the train or any other mechanical case, the composite speed would be very few different from the sum of the two speeds, and this as always explains why the effect would not have any practical significance.

The formula was derived both by Poincaré to complete Lorentz's theory and by Einstein in ED1905. But while within Lorentz's the formula has an obvious real meaning given the premise, in the case of relativity it has a controversial meaning as the whole theory. Indeed, given the arguing path with which relativity is deduced, the formula could have the legitimate meaning that the speeds of things observed through electromagnetic waves should be composed by applying the rule of composition instead of the simple sum, but it does not imply that the proper speeds of things cannot just be added through a sum. In the framework of relativity, the apparent speeds, that is, those measured by means of electromagnetic waves with instruments of the kind of a laser telemeter, should behave according to the formula, but in the premise of theory there is no reason to authorize saying that the real speeds cannot simply be added and in some cases be larger than c. But relativity introduces the formula with universal and realist meaning with the usual unjustified jumping over the premises and drowning in the imprecision of the text.

With these premises let us come to Fizeau's experiment, which dates back to 1851. The experiment was designed to resolve the dispute between two different theories about the ether that had been formulated and was realized by building an apparatus in which a beam emitted by a source is decomposed into two beams which traverse a pipe where water flows. A beam travels by running with the current and the other counter current. The scheme is this:

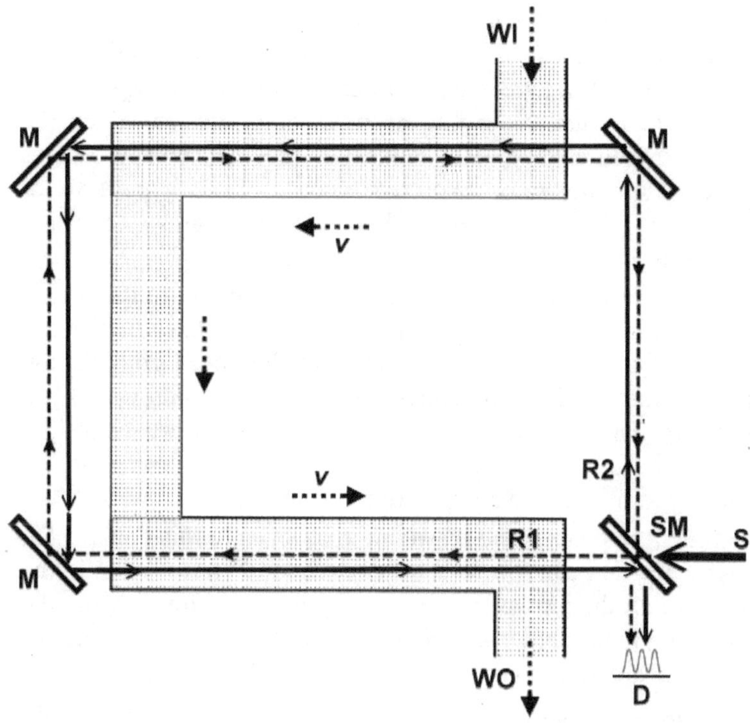

There is a transparent elbow pipe in which water enters through WI, flows at speed v and leaves through WO. A light beam is emitted by the source S and sent to the semi-reflective mirror SM reflecting part of the light while allowing another part to pass through it, and so splitting the initial beam into two R1 and R2 beams. The beams move in parallel and are reflected by the mirrors M. R1 rotates clockwise (dashed line), while R2 rotates counter clockwise (continuous line). Of course the mirrors are slightly inclined so that the two beams travel almost parallel without interfering with each other. At the end R1 runs through SM, R2 is reflected by it and the two beams are converged on a screen D where interference fringes are observed.

While in the M+M experiment the interference fringes depend only on approximations in the construction and are immobile, in this experiment of Fizeau the fringes manifest and vary according to the speed v of the water, which of course can be adjusted with different values, and can change sign if the sense of circulation is inverted.

311

The result of this experiment is therefore positive, not void as in the case of M+M. At Fizeau's time, it was found that the result of the experiment was not in agreement with any of the two interpretations of the ether that were taken into account, and between which the experiment should decide. The values of the interference fringes approximated roughly those expected according to a hypothesis of Fresnel dating back to 1818, but did not allow decisive interpretations.

In relativity we find this interpretation: by attributing to the beams the speed w that is that of the light in the water, the interference fringes would give a result consistent with the speed composition formula, and therefore the experiment would be an experimental test for the theory. That is, calculations show that the interference fringes would match the composed speeds of waves and water to be calculated according to the composition formula.

In the literature there is the reasoning that leads to this conclusion, and those who want can follow it for example in Ferraro[80]. But in the literature, not all the implicit and necessary premises for drawing the conclusion are highlighted, which are already evident given what we have said, and are numerous:

1) The speed composition formula applied to any experiment implies for itself that the contraction is interpreted as real and not apparent. We observed this before describing the Fizeau experiment and is not directly related to it.

2) Considering Fizeau's experiment, the conclusion that it would confirm the speed composition rule implicitly states that the dimensions of electromagnetic waves propagating in a fluid rather than in the vacuum are themselves subject to Lorentz's contraction. Be this true or false, this assertion is a tacit novelty introduced in theory.

3) The contraction specifically observed in this case is interpreted as real and not apparent: in fact the interference fringes are observed locally by an observer at zero speed with respect to them. In order to interpret the result in Einstein's relativity key, the phenomenon should be observed by electromagnetic waves that bear witness of it

---

[80] Ferraro 2007, Sections 2.7 and 3.6.

to an observer moving with respect to the apparatus. It is not easy to imagine how it could be done, but to be consistent with Einstein's premises one cannot draw conclusions from local observations. We must not be fooled by the fact that here we observe a phenomenon of electromagnetic waves: the phenomenon is not observed by means of other electromagnetic waves, but locally at zero speed, and this cannot give rise to relativistic consequence in the premises of Einstein. The misunderstanding is similar to that of the meson.

4) it is claimed that the speed of electromagnetic waves propagating in a fluid can be composed with that of the fluid. True or false, even this assertion is a tacit novelty introduced in theory. Do waves move with respect to the ether, with respect to the source, or when they cross a fluid do they move with respect to the fluid? The third case has never been heard before.

5) Finally, it is said that water and electromagnetic waves are contracted, while the terrestrial observer that measures the phenomenon is not. Symmetry is lost, as always when it comes to real consequences.

With all this, it is well-understood that the interpretation of Fizeau's experiment belongs with good right to the set of logical errors we have considered in the preceding sections, and consists of more than one. It may be that somebody in the future will draw important consequences by working on the old Fizeau's experiment, whose not null outcome is a pity to neglect, but until now it only serves to find out the inconsistency of relativity once more. It should be noted that while the experiment of the meson could be interpreted (by someone who wants to) as a confirmation of Lorentz's contraction with respect to ether, Fizeau's experiment cannot be, because the idea that the speed of waves is added to that of the fluid in which they move is strange to the ether theory: according to the hypothesis of the ether, the atoms of the ether are in the spaces between the nuclei and the electrons of the atoms of the fluid, the motion of the fluid takes place relative to the ether that remains at rest, and therefore the speed of the waves would remain relative to the ether, even if the waves are slowed down by the fluid crossing. In the ballistic case, it would be interesting to know how this experiment could be explained.

6.6   *The Hafele-Keating experiment and GPS system*

The Hafele-Keating experiment and the techniques used by the GPS system are considered empirical confirmations of special (as well as of general) relativity. To confirm what? The fact that time dilatation is apparent **and** time dilatation is real? The fact that the condition of uniform straight motion is irrelevant given the law of inertia **and** the relative speed is relevant to determine the time dilatation of a system rather than another? There can be no empirical verifications or falsifications of contradiction, because one cannot know which assertion would be confirmed or denied. Therefore, I do not comment in detail on these alleged experimental arguments: the idea of experimentally testing a recognizable contradiction as such is meaningless, and refutation would become nonsensical in turn. I am merely hinting at this to let the readers (if they want) the exercise to get documents and analyse the meaning of what the literature says about these issues in detail, with the premise that the purpose of discussing a contradiction can be just to look for motivations which lead men to the desire to contradict themselves and to blindness in front of the absurdities they claim.

The Hafele-Keating experiment of 1971, in which two atomic clocks were carried by plane around the Earth, demonstrated that atomic clocks taken out of the protected environment in which they usually operate are subject to frequency changes that are not corresponding to what is calculated according to relativity (because we do not know with respect to which frames the calculations should be developed), but determined by the various physical factors in the game.

As for the GPS system, referred to as the first technological application of relativity, the system is designed so that satellite atomic clocks synchronize again every few hours with stationary Earth clocks, because the environmental conditions in which satellites work in the presence of cosmic rays storms and other perturbations are a reason to take as standard the Earth clocks rather than those in orbit. Recent relativistic literature, on the other hand, asserts with certainty that the GPS system provides relativity application proofs, but it cannot give detailed accounts about it, because the detailed operation of the GPS system is still a well-guarded industrial secret by the US government. So the assertion is devoid of any documentation. I quote because it is funny what the Routledge Guide tells us, just the book which above[81] complimented us if we realized that dilatation is symmetrical[82]:

> Special Relativity tells us that because of the motion of the satellites, the moving clock will tick more slowly than clocks on the ground...

So we, who have invested so much effort to assimilate the law of inertia, now find that GPS satellites are moving while the Earth is stationary, perhaps at the centre of the Cosmos, and that therefore calculations of temporal dilatation apply to satellites and not to stations on the Earth. To talk seriously, the system's summary descriptions from official sources that are available make no mention of relativity, and a technician who was one of the major designers of the GPS system, Ronald Hatch, as a result of his experience came to deny any validity to Einstein's relativity. He wrote a book about and gave it the very clear title of *Escape from Einstein*, and this can be known by consulting an official US government website[83]. I cannot judge the book because it is difficult to find and I did not yet read it, but the author's professional curriculum suggests that it might be worthwhile to know.

---

[81] See above, Section 6.1.

[82] Trefil 2015, Chapter 8.

[83] http://www.gps.gov/governance/advisory/members/hatch/

6.7   Problem of the relativistic deduction of $E=mc^2$

A special problem of immense importance lies in the way the relativistic literature deals with the derivation of the formula $E=mc^2$ and the other equally important formula $m=m_0\gamma$. Let us resume the discussion from what we know, with the usual warning that what we shall say about this subject has the sole purpose of correctly framing the relationship between Einstein's relativity and the two formulas about mass, without proposing any solution to the problem inherent in this theme. To talk appropriately and in detail of this problem, nuclear and particle physics should be studied, and this is certainly not the appropriate place.

In Chapter 4, we saw that certain experimental premises allow us to derive those two formulas with a coherent chain of inferences: we used a merely didactic elementary derivation (Rohrlich) and a primordial derivation (Lewis), because we only need to understand how those formulas can be derived in a logically consistent way. For this purpose we used those sources and not the first derivation of Einstein of 1905, because the latter, which as we know develops the issue in only three pages, is just a sketch that applies the Lorentz coefficient to some radioactivity phenomena that were known and derives the formula $E=mc^2$ as an approximate result. For this reason, the first version of Einstein was then overlooked by all the literature, starting with Einstein himself.

Then in section 5.4 we saw some passages from a paper by Lewis and Tolman, that were useful to verify how in 1909 the interpretation of relativity was the conventional or apparent one (we know that the two cases for us are distinguished by a clear and unambiguous criterion, but it is not easy to distinguish them exactly in the literature) and how at that time the realist interpretation was explicitly excluded. No one thought that a clock compared locally with another could reveal a delay depending on the relative speed of a previous movement.

We also know that the two formulas $E=mc^2$ and $m=m_0\gamma$ imply an open problem: in order to be legitimate expressions of exact physical laws it is necessary that the speed of light c be a physical constant, and if Einstein's relativity is no solution, then this problem too cannot be considered solved today: the two formulas are legitimate only in the context of Lorentz's old hypothesis. In the 1908 paper Lewis had realized this, and in the last part of the paper he had problematically mentioned the subject.

However, in the hypothesis of Einstein's postulate that c would be constant for every observer, even if the propagation of light occurs with respect to the emitting body and not the ether, the two formulas do not need further justification for the use of the constant c: c becomes constant by definition, and although the assumption is bold and does not allow us to build any coherent kinematics, under this assumption the specific problem of mass formulas does not arise. To consolidate the marriage between the two formulas and the relativity theory which is necessary to make them legitimate, since 1909 relativistic literature performs an operation more royalist than the king, so to speak, and tries to derive the two formulas about mass from Einstein's premises without employing electromagnetic experimental data of any kind. This operation itself would not be necessary: if we have legitimized the use of c as a constant, then equations using c do not need to be included in the relativity building for its consequences to be legitimate: it is sufficient that c be a constant. However, the attempt is made, and it gives us an opportunity to highlight another typical feature and problem of the relativistic literature.

The derivation of the two formulas without the use of data supplying information about the mass concept appears in the 1909 paper by Lewis and Tolman, and then is always reproduced in similar form in the following literature; the 1909 paper seems to be actually the archetype. Probably the paternity of this paper should be attributed to Tolman rather than to Lewis, although it is signed by both, because Lewis later did not resume the topic of relativity, while Tolman was one of the most well-known advocates and popularisers of it throughout his life.

Considered that Einstein's 1905 derivation also employed electromagnetic considerations, not unlike the 1908 Lewis's paper, in 1909 Lewis and Tolman proposed this goal[84]:

The consequences which one of us [Lewis] has obtained from a simple

---

[84] Lewis and Tolman 1909, p. 713.

> assumption as to the mass of a beam of light, and the fundamental
> conservation laws of mass, energy, and momentum, Einstein has derived
> from the principle of relativity and the electromagnetic theory. We
> propose in this paper to show that these consequences may also be
> obtained merely from the conservation laws and the principle of
> relativity, without any reference to electromagnetics.

To achieve this, the two authors first construct the notion of space
and time dilatation with the usual technique, they derive Lorentz's
factor and explicitly state that the phenomenon is apparent, as we
have seen in section 5.4. Having set these premises, the authors begin
the derivation of $m=m_0\gamma$, which takes place in very short steps
through this argument[85]:

> Let us again consider two systems, a and b, in relative motion with the
> velocity v. An experimenter A on the first system constructs a ball of
> some rigid elastic material, with a volume of one cubic centimeter, and
> sets it in motion, with a velocity of one centimeter per second, towards
> the system b (in a direction perpendicular to the line of relative motion
> of the two systems). On the other system, an experimenter B constructs
> of the same material a similar ball with a volume of one cubic
> centimeter in his units, and imparts to it, also in his units, a velocity of
> one centimeter per second towards a. The experiment is so planned that
> the balls will collide and rebound over their original paths. Since the two
> systems are entirely symmetrical, it is evident by the principle of
> relativity, that the (algebraic) change in velocity of the first ball, as
> measured by A, is the same as the change in velocity of the other ball, as
> measured by B. This being the case, the observer A, considering himself
> at rest, concludes that the real change in velocity of the ball b is different
> from that of his own, for he remembers that while the unit of length is
> the same in this transverse direction in both systems, the unit of time is
> longer in the moving system.

Let us read carefully this first part of the demonstration (to interpret
the text, let us remember that according to the theory, in the running
system durations are contracted and clocks are slower, but time is
dilated and thus "the unit of time is longer in the moving system",
this clarification being aimed at avoiding the usual misunderstanding
between time and durations). We know that according to the theory,
if the motion takes place along the x-axis, then given the postulate of

---

[85] Lewis and Tolman 1909, p. 718.

the constancy of c for each observer the measurements of space along the x-axis and the time measurements appear modified according to the Lorentz factor, while space measurements along the y-axis do not appear to be altered because the speed along the y-axis is zero. Then, here it is said that "the real change in velocity of the ball b is different from that of a" because the transversal speed of the ball a is dy/dt, while the speed of the ball b is dy/dt'. It is not fully clear why the authors say "the real change", since in this context we would rather expect the expression "the apparent change", but the reason should be that here "real" means "calculated according to the theory". Given the premise of the theory, assuming that motion is on the x-axis, it is legitimate to formulate the following statements, not necessarily true but consistent with the premises:

- Distance measurements dy are common to the two systems ($dy=dy'$);
- From A's point of view, the time measurements of B must be corrected by applying the gamma factor: before the correction dt is different from dt';
- Therefore, the variations in speed on the y-axis direction require the same correction required by time measurements (those on the x-axis do not require it because space measurements are modified by the same factor as time);
- If the gamma factor is not applied, the change of the speed of the two balls on the y-axis appears to the observer A different for the two balls a and b.

But given these premises, what else can we draw beyond the conclusion that A must introduce the due corrections if he wants to know B's proper measures? Reading the following take into account that as always $\beta=v/c$[86]:

Velocity is measured in centimeters per second, and since the second is longer in the moving system, while the centimeter in the direction which we are considering is the same in both systems, the observer A, always using the units of his own system, concludes that the change in velocity of the ball b is smaller in the ratio $\sqrt{(1-\beta^2)}/1$ than the change in velocity of the ball a. The change in velocity of each ball multiplied by its mass

---

[86] Lewis and Tolman 1909, pp. 718-719.

gives its change in momentum. Now, from the law of conservation of momentum, A assumes that each ball experiences the same change in momentum, and therefore since he has already decided that the ball b has experienced a smaller change of velocity in the ratio $\sqrt{(1-\beta^2)}/1$, he must conclude that the mass of the ball in system b is greater than that of his own in the ratio $1/\sqrt{(1-\beta^2)}$.

In general, therefore, we must assume that the mass of a body increases with its velocity. We must bear in mind, however, as in all other cases, that the motion is determined with respect to some point *arbitrarily* chosen as a point of rest.

If m is the mass of a body in motion, and m_0 its mass at rest, we have:

$$\frac{m}{m_0} = \frac{1}{\sqrt{1 - \beta^2}}.$$

And so the formula $m = m_0 \gamma$ was derived using only the principle of conservation of momentum and no experimental information. Subsequently this argument technique leads to the derivation of $E = mc^2$ similarly, and this derivation technique will be resumed by the literature in a similar manner in each treatise, beginning with Einstein in the popular account of relativity. It is not necessary to scrutinise the follow-up in detail, because what we have read already puts us in front of a set of problems without solution, as there is no solution to the problem of the decision between apparent and realist interpretation of relativity in general. Those who want to deepen their understanding of the issue by reading the derivations of $E = mc^2$ that are found in the literature, will find that they are based on the imaginative application of the γ factor to momentum without any reason that authorizes us to do so.

Before commenting on Lewis and Tolman's text, let us make a general remark. We have three fundamental quantities of mechanics, which are space, time, and mass. So, as long as we continue to conceive these three quantities as primitive concepts heterogeneous among themselves, or even if we reduce them to two by distinguishing only space-time on one side and mass on the other, if we set up a reasoning with premises that will give us information about space and time, but not about mass and not even about energy (which is not a primitive concept, but is defined from the other three fundamental concepts), then we cannot derive information about the concept of mass from a reasoning set up this way. I see a distant ship and wonder: how long is it and how much does it weigh? If I make measurements with instruments and find, for example, the speed of the ship's passage and its distance to some coastal reference whose measurements I know, I can conclude that the ship e.g. is two hundred metres long. But only from these information regarding merely lengths and durations, I can never deduce how much the ship weighs.

Returning to the argument of Lewis and Tolman, whose premises, according to the explicit author's intent, contain no information on energy and mass, let us ask ourselves: does the argument speak of an apparent or of a real mass increase? Given all their premises, and the way Lewis and Tolman have described relativity, and the logical impossibility of inferring anything regarding the mass concept in the absence of information in the premise, we conclude that this is an apparent increase. And so in order for the increase to have physical meaning, we should pass with an unjustified leap from the apparent interpretation to the realist one as ED1905 does to get the result of the invariance of Maxwell's laws as a result of apparent time dilatation. Without this leap, applying the theory we could say only: a mass moving against me at speed v appears to me at a higher value than the m_0 value that would be measured at zero speed, but this only applies to remote measurements with electromagnetic signals which I could perform without taking into account the necessary correction factor; if the mass collides with me, the kinetic energy of the impact is the one that is determined given the speed v and given m_0, not given m incremented as a function of the speed v. And after having interpreted things this way, there would be no explication for the experimental data attesting that in accelerators the electric power required to increase the speed of the particles has a trend corresponding to the Lorentz factor. So suppose we make a logical leap and interpret Lewis and Tolman's arguments by asserting that the mass increase with speed is real: this would be an arbitrary leap, because inside Lewis and Tolman's argument there is nothing that authorizes us to do so. And the logical leap would immediately lead us to the same problem of clocks: since everything in the world moves at infinite different speeds with respect to infinite different frames, everything would have endless relative masses, and mass would cease to be an objective and invariant property. In the framework of the ether, we could very well conceive that the mass increases according to the speed with respect to the unique reference represented by the ether as well as we could conceive that a clock lags in consequence and as a function of speed with respect to the ether. The same could be conceived with respect to any other universal reference to motion, if one existed. But in the framework of relativity and universal extension of the law of inertia, this possibility is logically denied. And so the issue of mass growth in nuclear and particle physics would need today or tomorrow a personality of a temper like Newton who could re-interpret critically all the acquired knowledge. Here nothing can be said about it.

7.   *Success of special relativity and opinions*

7.1   *Success of the theory*

The story of the immense success of relativity has been told many times in hagiographic terms by authors who were or are unable to realize the absurdity of theory by breaking the spell of suggestion. Special relativity circulated among theoretical physicists after 1905, stirring up the enthusiasm we have seen in Tolman and Langevin and others. Einstein published general relativity some years later, and in 1919 some astronomical observations that gave controversial confirmations of predictions of general relativity unleashed literally overnight the explosion of Einstein's media celebrity, which has lasted until the present. The story is told very well by Isaacson[87] among others. The reason for this media celebrity is obscure: the phenomenon was triggered by the fact that according to calculations of general relativity, the gravitational effect of the Sun on a light beam would be different from that calculated according to classical mechanics. Even classic mechanics predicts that electromagnetic waves are subject to gravity in the hypothesis that they have mass or momentum, but according to general relativity the amount of the phenomenon would be slightly different from that calculated according to classical mechanics, and some astronomical observations then considered accurate would confirm[88] Einstein's calculations. The reason why all this exerts so a strong influence on everyone's opinion is mysterious. But the fact is that there was an explosion of interest in the press: in 1919 Eddington announced the results of those observations, and the day after the news was on the front page of *The Times* in London; three days after the news reached the first page of *The New York Times* and then the rest of the world press. This was how the news was presented in the *New York Times*

---

[87] Isaacson 2007, Chapters 11 and 12.

[88] See also just how Isaacson 2007 speaks about this in Chapter 11 to get an idea of the accuracy of these measures. Stephen Hawking himself admitted that these experiments were of little value: see McCausland 2011, p. 209.

on the second day on which this newspaper put the story on the front page[89]:

LIGHTS ALL ASKEW IN THE HEAVENS

Men of Science More or Less Agog Over Results of Eclipse Observations.

EINSTEIN THEORY TRIUMPHS

Stars Not Where They Seemed or Were Calculated to be, but Nobody Need Worry.

A BOOK FOR 12 WISE MEN

No More in All the World Could Comprehend it, Said Einstein When His Daring Publishers Accepted it.

How and why such a specialist detail about the light of stars in the sky determined the sudden birth of a pop star on the Earth is a fascinating problem that deserves investigation, but in the next pages we shall only deal with another even stranger phenomenon than general public credulity: that of consent to special relativity by specialists and well informed people, i.e. by those who would have full capacity to realize that the theory was nothing but an attempt to solve a certain problem with the least effort through a senseless conventional assumption.

We have seen that understanding of special relativity consists of entering into a singular state of consciousness in which a person agrees to a set of propositions that can be reduced to this general scheme:

"there is a phenomenon called time dilatation that is apparent and therefore can be and is equal and symmetric between two systems A and B"

and

"given this phenomenon, A's clocks appear late to B and B's clocks appear late to A"

and

"given this phenomenon, A's clocks are delayed in a not apparent way with respect to B's clocks."

---

[89] Text reproduced in Isaacson 2007, Chapter 12.

The fourth proposition, which should be pronounced by symmetry, "and given this phenomenon, B's clocks are delayed in a not apparent way with respect to A's clocks", is never pronounced but is exorcised and kept in the shade through rhetorical gimmicks. There are a few variants, but the immense amount of written words revolves around a small extension theory that expands quantitatively in the rhetorical apparatus built upon it, and then the process is reproduced likewise in general relativity, which depends on the special one and inherits its method. We did not talk about general relativity, but we have had a small assay of how it proceeds in section 5.6: general relativity is based on the description of some phenomena initially declared apparent and then treated as non-apparent properties of physical reality on basis of the unjustified interpretation of some mathematical consequences.

The phenomenon of consent to special relativity is one of the strangest and most singular human events that we can find in history: a community boasting of being the first heir and the ultimate incarnation of rational thought decides to enter into a state of cognitive dissonance and carries out self-deceiving strategies to adopt a theory that is meaningless because it is contradictory. Each theory has its hypothetical premises and therefore is subject to evolution, but here we have a non-theory that lacks the basic requirement of coherence. There is no myth that has such a radical quality: every myth and every fairy tale paint situations conceived to give satisfaction to the needs and desires of the community that identifies itself through them, but always on the base of fantastic representations that in theory could even have reality, or have a tiny but not insignificant chance of happening. Once Festinger[90] studied the phenomenon of cognitive dissonance by infiltrating his researchers into a community that firmly believed that on a certain day in 1954 the Earth would be shocked by cataclysms, but the good people would be saved by flying saucers sent by the same superior minds who had decreed the cataclysms to punish the wicked. Cognitive dissonance manifested itself when nothing happened and the trusting believers' community had to find a way to rationalize the experimental falsification of its predictions, but there was never

---

[90] Festinger 1956.

literally a logical contradiction in the assertions of these people. They wanted to assert the reality of a picture of events that of course was the result of imagination and puerile expectations, and therefore, in the face of experimental disproof, they attempted to rationalize it and to find reasons for the changed decisions of the superhuman powers that had changed their plans at the last minute for good purposes, or if they had deceived the good believers, they had done so well to test and strengthen their faith. In all this, the key is not contradiction, but it is the obstinate will to attribute the status of reality to fantasies produced by human fragility. When this pertinacious will manifests itself, it always stems from the subjective inability to appropriate the culture of the modern world: this is what happens in all subcultural phenomena.

Relativity offers a much more disconcerting experience: science-equipped people assert the set of contradictory assertions we know, and thus form a privileged community to which they have access by accepting the state of dissonance and blindness with respect to the contradiction they all share. The process is favoured by the fact that to bring to full consciousness the contradictory scope of the assertions of relativity, it is necessary to analyze them thoroughly: those who are superficially listening are vulnerable to appropriate the theory in a merely imitative way, without entering exactly into the state of dissonance, just repeating passively statements having no sense for them. Many do not contradict themselves, but only repeat words by applying the liturgical formulas that are learned from literature. Some, however, contradict themselves intimately: someone has to write the new textbooks on relativity, and some must have originally invented the system of absurd assertions that make up the rhetorical apparatus of theory. And the greatest paradox is that those who identify themselves more intimately with the theory and its contradictions are exactly the ones who know more, and should be more equipped to exert the criticism we have made in this book and that all in all is not that difficult. On the contrary, the ignorant and inexperienced who approach relativity with the candour of the novices mostly perceive the absurdity, and then given the prestige of the theory they attribute to their own limits the fact that they do not understand it, but do not enter the state of cognitive dissonance. Of course, this honours the intelligence of the many who do not understand.

The anthropological and cultural history of relativity today is to be written: it was never written for the obvious reason that anyone who would be able to write it usually is unable to understand critically relativity, and therefore does not realize the problem. The public believes in justified good faith that the theory is something solid and so complex that only experts at the highest level are able to understand it, and we all have gone through this belief. Attentive analysis shows that things are very different: the elite of theoretical physicists have behaved irrationally in relation to this problem for a hundred years, which marks a radical divorce between experimental science and technology on the one hand, and the self-sufficient theoretical speculation on the other. In this book we have quoted the passages of some early relativists, Tolman and Langevin, to begin reconstructing the history of consensus on relativity. What we have seen is useful to better understand the meaning of the theory, and also gives us an indication of a fact that with all that we know at this point is an easy consequence and not subject to doubt: that the author of relativity is not Einstein as an individual, but it is a generational community, that of young physicists shortly after 1900, who saw in Einstein's theory the sketch of something that was far beyond the unresolved scientific problem. Through his disconnected thoughts of ED1905 Einstein provided the opportunity to begin a game whose stake was not the invariance of Maxwell's laws; it was the assertion that nature would violate the logical requirement of the unity of time. The overwhelming enthusiasm for this idea was the obsession of the generation around the Great War, and the old experts were overshadowed; nowadays it would be necessary to study in depth the context of Einstein supernova's birth between 1905 and 1920, but what is said in the hagiographic stories in circulation gives an already clear picture of the attitude of old-fashioned physicists. Max Planck and Lorentz, who were both in personal relationship with Einstein, left perplexing testimonies in which they expressed an acceptance of the theory with unclear reservations. In the 1916 edition of Lorentz's work on the electron there is a note missing in the previous edition of 1909[91]:

> If I had to write the last chapter now, I should certainly have given a more prominent place to Einstein's theory of relativity by which the theory of electromagnetic phenomena in moving systems gains a simplicity that I had not been able to attain. The chief cause of my failure was my clinging to the idea that the variable t only can be considered as the true time and that my local time t' must be regarded as no more than an auxiliary mathematical quantity. In Einstein's theory, on the contrary, t' plays the same part as t; if we want to describe phenomena in terms of x', y', z', t' we must work with these variables exactly as we could do with x, y, z, t.

This is a perplexed observation that also puzzles the reader, because either Lorentz recognizes that Einstein's theory, which is certainly more elegant than his, is also sustainable, or he thinks it is as elegant as anyone wants, but not sustainable. Since Einstein's version is flawed by a fundamental logical deficit, the two theories could not be placed side by side as if they both had plausibility reasons, but Lorentz himself seems so much dragged by the current that he does not notice it, or does not want to notice. Michelson unsuccessfully opposed the theory; Poincaré, who died in 1912, avoided talking about it, but after 1905 he gave less emphasis and was inclined to resize the scope of his principle of relativity to which he had previously given the utmost importance. Rutherford is reported to react to relativity with irony, but nothing more than this: it does not seem that he attempted to stem the phenomenon of suggestion he saw around himself. The old people seem to have been paralyzed by the explosion of the generational enthusiasm phenomenon against which rational arguments could nothing, as the arguments of rational pacifism could do nothing against the 1914 bellicose fury of the whole European society.

But why did the violation of the unity of the time exert and why does it still exert such a powerful fascination? Why, in order to support this assertion, did a generation expose itself to the absurd, which implies, among other things, a state of supreme indifference and lack of curiosity for nature itself? Why from this was born an

---

[91] Lorentz 1916, p. 321.

academic tradition that has continued for a hundred years to repeat the same absurdities, which, moreover, with the passing of time consolidate themselves better, because to ignorant and novices it seem less and less likely that relativity may be wrong? How could it be possible, say people who do not know the theory, that in a hundred years it has never been denied? And therefore the theory is strengthened in a self-feeding process, and it will continue to grow until some great scientific minds in the future will revolt against academic nonsense with a decisive rupture.

As for the fascination of the theory, we might ask ourselves this: Lorentz's hypothesis also describes a condition in which clocks would be discordant and in which simultaneous events would correspond to different time values in two moving systems. In addition, Lorentz's hypothesis speaks of everything in plausible terms, though not confirmed by any experience. Yet no one has invested in Lorentz's hypothesis, giving it the symbolic value that was given to Einstein's version of the same hypothesis. Why does Lorentz's hypothesis not fascinate and the other does? We can suspect that fascination originates precisely from the contradictory way of presenting itself, which is the feature of Einstein's version.

7.2   *(Perhaps) naive considerations about the unity of time*

To find out what might be the root of enthusiasm for relativity in Einstein's version, let us allow ourselves some old-fashioned and naive considerations about the unity of time in the human vision of things, with the warning that these are considerations of a kind that anti-classic physics of the twentieth century considers barbaric.

As we know, Einstein worked out the idea that two simultaneous events are known when two remote systems exchange a signal, and therefore that simultaneity is connected to the signal speed; the roots of this lay in Lorentz's and Poincaré's work, and in the observation of the mechanism of some early twentieth-century inventions to electronically synchronize two or more clocks. Then Einstein added to this the assertion that this mechanism would supply the primary and fundamental definition of simultaneity, and argued that if we postulate that the speed of light is constant in all inertial systems, the judgment of the simultaneity of two events and the synchronizing two clocks necessarily takes a certain form. On these bases we cannot build a kinematics without incurring crucial contradictions, but now let us neglect this issue.

This view on the issue of time is always opposed to Newton's old idea, which postulated the existence of a single time containing all things, described as a straight line and having geometric properties identical to those of the straight line in Euclidean geometry. Einstein's methodological argument is this: we can attribute a determined sense to a procedure by which we exchange signals and declare simultaneous two events. This procedure is usable and used in science and technology. On the contrary, Newton's unique time is a construction of the imagination, and we cannot do any experimental testing concerning it. So the definition of simultaneity as a phenomenon that occurs in the exchange of signals contains all the theory of time that science needs.

Let us see, however, what is our experience in a wide human sense: how do we behave when we perform any determination of time? We know and we verify in all our cognitive experiences that when we think of a relationship between two things, when we establish a theory describing a relationship between any pair of systems, we can only proceed by putting together the development of the two systems in a single time. If we think of two trains running on two different lines, we can conceive them both in their own time if we want, and until we are not looking for any relationship between the two journeys we actually have two different times: the judgment that the two journeys take place either one after the other, or simultaneously, or in partial simultaneity, is absent from our consciousness or is arbitrary until we decide to gather the objective information necessary to pronounce it. The same happens, for example, if we think of two plants growing one inside a greenhouse and the other outdoors. The first one blooms in February, the second matures more slowly and blooms in April. If we look at both of them as isolated systems without thinking of any relationship between them and with anything outside of them, we see that the same identical biological process takes place over time, but each plant does so in its time that is neither dilated with respect to the time of the other, nor contracted nor identical, but is devoid of any relationship. But if we ask ourselves any question about the relationship between the two systems, then we soon put the times of the two systems together by conceiving them as segments parallel to a single line of time. So in our example we would say that the maturation of the plant in the greenhouse is quicker than the other, and the time measures of the two processes are no more identical, although their biological aspects are. If we do not do this, we cannot think of any relationship between two systems, and this corresponds to a logical need that is as fundamental as it is the need for non-contradiction or for compliance with the basic rules of arithmetic.

Let us go back to the example about two trains: some people are travelling on a train from A to B, and they ask: shall we arrive in time to catch the train from B to C? To answer, we must represent the two durations as parallel to the unique line of time: there is no other way of setting the problem and giving meaning to the words that express it. Then if the situation is this:

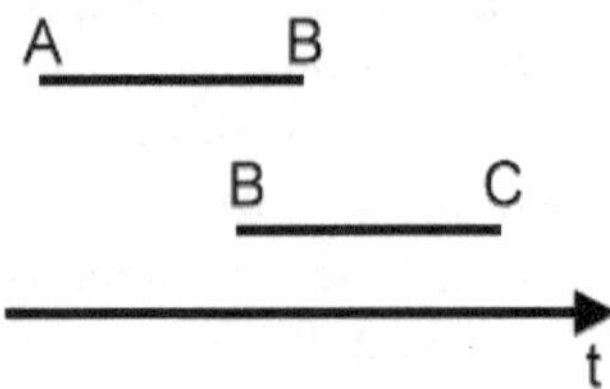

the answer is, "no, we shall not have time to get on the train BC", because the train BC leaves B before the end of trip of the train AB, so we shall arrive at Station B when the train with destination C will be gone. But if the situation is this:

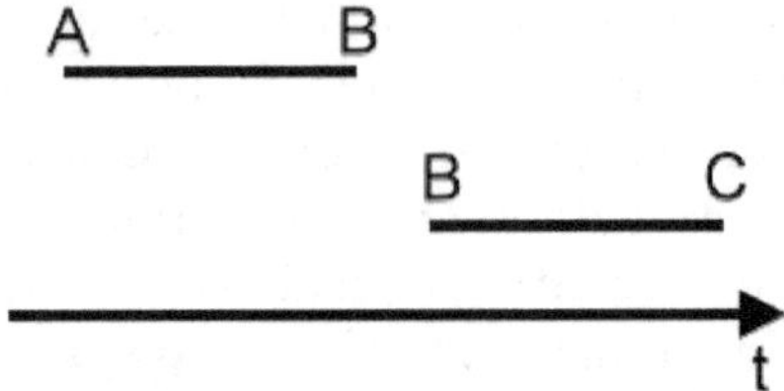

then the answer is: "yes, we shall arrive in time" because the train BC starts after the end of the trip of the train AB.

A problem can be as complex as you want, but in any problem the logical mechanism of temporal determination is reduced to this elementary scheme, which is the basic condition to give a meaning to the more or less articulated techniques that are needed to measure the duration or to decide about the succession or simultaneity of phenomena. We can well synchronize our clocks with electromagnetic signals: but if we are not able to first draw the geometric time according to the pattern in our example, then we would not make any synchronization, because the synchronization operation would have no meaning for us. For the same reason, people would not endure fatigue nor would they be able to count the banknotes they received in payment in exchange for something, if they did not know the property of numbers before doing so. For this reason, defining time through the procedures that can be conceived to measure it rather than through its elementary geometric representation is such a foolish operation as it would be to imagine an animal which receives a hundred banknotes and without knowing anything about numbers counts them thinking of the quantity of pet food cans that could be bought. First comes the ability to think in geometric and arithmetical terms, and then come the techniques that implement calculations of increasing complexity and not vice versa. First comes the geometric representation of time, and later come solutions for timing problems. Anyone who thinks the opposite, for coherence should abandon any concept of necessity of mathematical demonstrations and profess even a radical mathematical empiricism.

Does the fact that the unitary and geometric conception of time precedes the measurement techniques imply that there is only one absolute time, as described by Newton? The consequence is not this. There could certainly be worlds having each one its own time: we cannot deny this possibility in abstract, and we cannot say anything about this possibility. What we find in every experience, is that we can form a concept of any relationship between things only if things can be built geometrically within the only time that we are able to conceive. The human mind possesses a time, no more than one, in which it conceives phenomena that take place with multiple and parallel durations. The fundamental time in which everything develops its phenomena does not have any physical existence, because it is only a reference system, and it is neither finite nor infinite because it is nothing more than mathematics. As it does not make sense to wonder if there is a number ending the integer series, so it makes no sense to wonder if there is a start or an end of time. What we know is that given a number n, we can always say: $n+1$ and $-n-1$, and this assertion for us has a perfectly determined sense. It is impossible to conceive the last positive or negative number. Equally, what we know is that given a time t we can always say: $t+1$ and $-t-1$ and this assertion for us has a perfectly determined sense. What is impossible to say is if there is a time t after which there is no more any real thing, or if there is a time $-t$ before which there is no more any real thing: this question is devoid of sense because geometric time is no physical reality, but it is a mathematical form through which we think on physical reality. So the insoluble question "if there is a time t after which there is nothing left" should be expressed in the present tense and not the future: time is mathematics, where there is no past and future. Only the events that populate time are past and future inside the consciousness that receives them, and therefore among the many issues we cannot solve there is the question whether the distinction between past and future is real or illusory, if it really is in things or whether it is just a property of the way things are presented to the human mind.

People are able to know only those things that are presented as experience so that they can be measured in relation to a single time axis. Therefore, we know of nature only what can be conceived within the unity of time. Things or worlds that could exist in a time different from ours would not only never appear to our intuitive perception, but they would not even exist in our thoughts. That there are worlds in their own time distinct from ours is a hypothesis that we cannot deny, but we cannot even determine it: we can only enunciate it. Therefore, we have no reason to confuse the concept of time and simultaneity with the procedures by which we can ascertain simultaneity. The concept of simultaneity is purely mathematical, and consists of saying: on the axis of time two events have the same value t. Instead, the procedures for determining time coordinates, measuring time, and pronouncing judgments about simultaneity or succession of events depend on perceptional factors, technical procedures, and knowledge of the physical features of things that are judged, and therefore have no absolute value. In the example of travellers on a train AB wondering if they can catch the train BC, one of the answers could be this one:

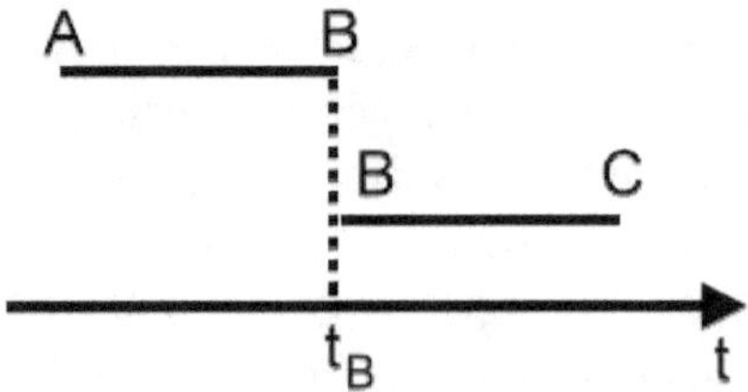

"Who can tell if I can catch the train BC if the train AB arrives in B just as BC goes, how many seconds will be available to switch from one train to another, how far apart will the two tracks be?" The simultaneous occurrence of BC's arrival and BC's departure from a station as a purely mathematical problem is perfectly and uniquely defined by the abstract geometric representation, and its meaning is clear. The problem must be represented in a geometric scheme like this if we want the words to have meaning, and to understand the problem to be solved it is enough we understand the meaning of the expression: "two events occur at the time that in this example is called t_B". Then there are specific time measurement problems to solve, and here we encounter different complexities according to the cases: there are cases when we just have to look at the position of the sun or at a clock dial, and others in which we need to build complex machines which allow us to associate events with the ticks of atomic clocks running at very high rates. In any case, with respect to any problem we need first to conceive the problem of simultaneity in the purely formal representation of the unique time and then define procedures to ascertain simultaneity, which can be elementary when the perceptional brain apparatus is capable to pay attention to two things in the same state of consciousness (I am getting into the station and I see that the other train is already leaving, or I see that it is not yet moving and people are still climbing on it), or may require the invention of procedures not elementary and anything but intuitive, as happens in any measurement that requires the invention of a technology.

These considerations are barbaric according to twentieth century theoretical physics, but are also perfectly matched to what always happens everywhere scientific results are produced, including experimental science and technology in the twentieth century and nowadays. One might object, however, that the classic conception of the unity of time is an opinion and a hypothesis that is as valid as the conception that admits multiple times. Instead, it is not so, and the very hypotheses we know, the consistent one of Lorentz and the incoherent one of Einstein, both show how the unity of time in the end has the last word. In fact, asserting the unity of time, it is not equivalent to state that multiple times cannot exist even hypothetically, but it is to state that the processes of scientific object construction of men always lead to reconstructing the unity of time in the phenomenal world that can be built in concrete by human knowledge. This happens in both theories: both do not talk about unrelated times, but give the key to calculate each time as a function of the other, each t' as a function of a t value.

This observation now offers us a first explanation, not complete, of the fact that the fascination with relativity comes from Einstein's version and not Lorentz's. If we look at things without stopping at the surface, we realize that the definition and form of Lorentz's transformations does not deny, but confirms the need to put things in the formal unity of a single mathematical time. It is conceivable that there are worlds or systems where all durations are different than those of other systems and where simultaneous events are phase-out. This is perfectly clear in the context of Lorentz's contraction, and in Einstein's hypothesis this thing does not become inconceivable for itself, but is only incompatible with the law of inertia and it is impossible to measure it as a function of speed. But if there were multiple systems in which clocks and any other phenomena were moving in time at different rates, these systems would not know anything about each other until they for some reason decided to study their correlation: nobody denies that. But as soon as we wanted to know the relationship between two systems, we would refer their temporal coordinates to a common time frame, chosen arbitrarily, and we would try to measure the relationship between the two systems, for example with Lorentz's formulas or some other function. Immediately the local and multiple times would disappear, and we would simply say that inside the system A all phenomena are slower than in system B: this is exactly what happens when Lorentz's transformations are introduced, and is what would happen in Einstein's version of relativity too, if it had such features that it could be used to determine consistently any phenomenon, empirically ascertained or even just hypothetical. Thus, the relativity in both versions ultimately proves the unity of time: to think anything about the relationship between the times of two systems, a transformation formula is sought because else we would not know what else to look for, and if there is a formula to correlate the durations in the two systems, as a consequence we could no longer say that the times are distinct, but we would simply go back to say that the phenomena of a system are all faster or slower than those of another according to a precise proportionality coefficient.

But with this we also understand that Einstein's version just because it is incapable of generating a coherent picture of things can keep alive the purely verbal assertion of the multiplicity of times. The central statement of special relativity is: "there is no absolute time, but every inertial system has its own time, depending on its speed and as a result of signals' features." What does this statement mean? What representation arises in the mind of the recipient? The reader of this book perfectly understands the semantics of this assertion: it signifies the trial to solve the problem of Maxwell's laws invariance without reformulating electromagnetic theory, and alludes to a way to achieve this goal that would be ingenious, but that is impossible to follow. There is no further meaning, but only the statement of a working hypothesis that is shown impossible if thorough logical verification is developed.

Instead, for the twentieth-century world, the assertion "there is no absolute time, but every inertial system has its own time, depending on its speed and as a result of signals' features" evokes extra-scientific and symbolic promises that are related with deep and still mysterious expectations. Books are published where experts at the highest level describe unusual and picturesque spaces and times, and the general public reads them avidly with the utopia to become able to master the deductive processes that lead to those visions, while believing everything is read. One might object: but it is not said that everything that is written about Einstein and everything following him should be as inconsistent as special relativity. This could be true. However, the authors describing space-time scenarios in the twentieth-century physics style, very different from the tradition of experimental science, are all authors who regard relativity as the ancestor of everything following and are immersed in the state of dissonance with respect to the absurdity of special relativity. Therefore those who have the gift of understanding what they read and being aware of the contradictions they find expressed, have the right to be suspicious, and to say, where there is a clue of the method of relativity: I do not believe anything if I myself cannot fully test the deduction chain.

The point that generates the different symbolic value of relativity between Lorentz's and Einstein's is this: the rationality of Lorentz's hypothesis first goes through the concept of multiplicity of times, but then leaves it behind by solving the problem with the transformations that correlate all the measurements among themselves through the common reference: the theory is still fully classical. Einstein's version, on the other hand, proposes the same mathematics, but leaves every problem indeterminate and insoluble due to its contradiction. In this way, in Einstein's version, the fantastic representation of the multiplicity of times is preserved untouched and pure, and although it has no definite meaning, it retains its gratifying character for fantasy and gets the qualification of being beyond the limits of classical thought. This helps us to understand why the emotional investment was made for Einstein's and not for Lorentz's: Lorentz leaves no room for irrational symbolic expectations. But the explanation is partial, because the reason why the assertion of the multiplicity of times is so full of symbolic extra-scientific meanings and why it became a fetish of our world remains still completely mysterious.

7.3 *Connections with twentieth century culture*

Readers will say: but it is quite clear why the assertion of the multiplicity of times has assumed such a deep symbolic charge. The era of this process was the same in which Bergson's philosophy taught that the time of consciousness is not geometric as that of scientific object construction, but grows by accumulating the remembrances that are present in consciousness, and was the same era in which Proust wrote his *Recherche* where the memory of the most minute events returns the universal meanings of his world. Certainly, relativity is connected with the culture of the era before 1914 and with that of the subsequent times according to the multiple threads of a story that has yet to be written. Perhaps the Great War has changed the sign of this relationship: it seems sometimes that the anti-war era was aggressive and wanted to kill something unbearable in nineteenth-century culture, while the post-war period perhaps looked for a sort of protective figure in Einstein's icon. Sooner or later we shall have a clearer idea about. Only, this has no significance or knowledge value about the physical world, and therefore relativity appears to be a poor thing compared to Bergson's philosophy, Proust's art, and the whole culture of the early twentieth century: relativity does not actually deal with physical world, but pretends to deal with the spiritual, and does so like an amateur who speaks metaphorically and uses in a misunderstood way the little which is known of the physical world. The presumptuous approximation of nineteenth-century positivism continues to triumph in a certain sense, and the complacency for the fascination of multiple times when they appear to have no logical base looks like an amateur velleity when compared with the penetration capacity of the deep conflicts inside the human soul that the culture of the early twentieth century had.

The cultural history of relativity is to be written, and it will be written on the condition that those who attempt it see with Cartesian clearness the physical nullity of the theory. In the absence of such clearness, the cultural significance of relativity was acutely glimpsed by many, but the underlying insight could not be developed, as a result of the belief in good faith in the consistency of the theory. This is exemplary in this brilliant passage by José Ortega y Gasset[92]:

> The provincial spirit has always been, and with full reason, considered a clumsiness. It consists of an optical error. The provincial does not realize that he looks at the world from an eccentric position. He assumes, on the contrary, that he is at the centre of the globe, and judges everything as if his vision were central. Hence a deplorable sufficiency that produces such comical effects. All his opinions are born falsified, because they start from a pseudocentre. On the other hand, the man from the capital knows that his city, no matter how big, is only a point of the cosmos, an eccentric corner. He also knows that there is no centre in the world and that it is therefore necessary to discount in all our judgments the peculiar perspective that reality offers when seen from our point of view. For this reason, to the provincial the neighbour of the big city always seems sceptical, when he is only better informed.
>
> Einstein's theory came to reveal that modern science, in its exemplary discipline — Galileo's *nuova scienza*, the glorious physics of the Western World — suffered from acute provincialism.

---

[92] Ortega y Gasset 1923, Appendix: "El sentido histórico de la teoria de Einstein". Text: "El espíritu provinciano ha sido siempre, y con plena razón, considerado como una torpeza. Consiste en un error de óptica. El provinciano no cae en la cuenta de que mira el mundo desde una posición excéntrica. Supone, por el contrario, que está en el centro del orbe, y juzga de todo como sí su visión fuese central. De aquí una deplorable suficiencia que produce efectos tan cómicos. Todas sus opiniones nacen falsificadas, porque parten de un pseudocentro. En cambio, el hombre de la capital sabe que su ciudad, por grande que sea, es sólo un punto del cosmos, un rincón excéntrico. Sabe, además, que en el mundo no hay centro y que es, por tanto, necesario descontar en todos nuestros juicios la peculiar perspectiva que la realidad ofrece mirada desde nuestro punto de vista. Por este motivo, al provinciano el vecino de la gran ciudad parece siempre escéptico, cuando sólo es más avisado.

La teoría de Einstein ha venido a revelar que la ciencia moderna, en su disciplina ejemplar — la *nuova scienza* de Galileo, la gloriosa física de Occidente —, padecía un agudo provincianismo."

These are evocative considerations which are also difficult or impossible not to share, until it is thought that relativity is an authoritative theory. Ortega y Gasset opened his own essay with the premise that Einstein's theory, being a theory by definition could need correction or even be erroneous, and pointing out that he was only concerned with the cultural process that had led to the "bloom of the plant", irrespective of whether the fruit was "healthy or harmful". But what would Ortega y Gasset have said if he could see that the theory from which it all came was not a questionable assumption, but an immature and conventional attempt to solve a problem that remains unresolved, and which specialists should easily recognize as such? Would he write in these same terms? Or would he not reverse the judgment, and rather investigate the meaning of the authentic provincialism disease that affects specialists in relativity?

Instead, the presumed and apparent provincialism of those who realize that the relativity theory is inconsistent, but are not able to fully express their perplexity, nourishes the sense of superiority of those who enter the state of dissonance that allows them to profess and repeat the theory, and receive the rich prize which is at stake: inclusion in one of the most esteemed social elites, if not the most esteemed overall. The authors of relativistic treatises often express themselves with insolence towards their readers, who are imagined in difficulties as they face paradoxes: obviously for authors those who do not understand the theory are not clever enough and are not up to modernity; seeing paradoxes means being captive of "remnants of classical thought"[93]. It would not be worth remarking this peculiar issue, if even this were not in the end a unique feature of relativity: those who buy a textbook on relativity, and hence are the customer, pay to be ridiculed by the author; the game has an obvious sadistic and masochistic character, and at stake is the construction of a hierarchy of domination and subordination. In the general case the opposite happens: theorists often support erroneous views, but are not oppressed by the problem of reducing the inner discomfort of the dissonance of a theory that ultimately they know to be contradictory; so they satirise other authors with whom they are in contrast, but try

---

[93] Literally quoted from one of the many textbooks on relativity we consulted.

to create a spirit of community with their readers, and have respect for those who invest their time in listening to them.

7.4   Rational criticism of relativity

One can guess that rational critics of relativity have been an unlucky minority, and that is right: rational critics have paid a price for their positions against the academic mainstream, but they have left witness of their thinking process, and their stories are interesting, and will be very useful both to those who will return to the conceptual problems left unresolved and to those who will write without credulity the cultural history of relativity.

In addition to the small legion of authoritative and rational critics, there are critics of relativity of two other species. The first is represented by anti-Semites, neo-Nazis, conspiracy believers, and the like. Browsing the web for a few minutes one can find as many as one wants. These are people who do not deserve discussion. The second is represented by the naive or mavericks, and many of these too can be identified quickly on the Internet. Many of the naive are pseudoscience practitioners, but not all. The problem with the naive is that sometimes they see the point, but instead of focusing on the state of things and trying to create a clear perspective in relation to the problem of relativity's inconsistency, they hurry to expose their own alternative theories to relativistic ones about electromagnetism, gravity, cosmology, and more. In this way, they reveal the recklessness of dilettantism, which sometimes might have less illogical consequences than in Einstein's case, but which has no hope to be forgiven and taken seriously. Anyone who wants to say a new word in relation to unresolved physical problems should present their own research with humility and caution that suits the theories concerning controversial matters, and to separate their proposal from critics to the contradictory condition of relativity. Instead, the naive and amateurs are struggling to solve the problems left open, and so they not only miss the opportunity to be taken seriously, but they prove that the bad lesson from the relativity method has caused damages that involve even those who would retain some ability to see the inconsistency of the theory professed by the academic mainstream. This applies to those of the amateur critics who realize the contradiction of relativity; many others profess opposite conclusions on the basis of the same methods of relativity, and therefore their work is fated to be ignored by everyone because orthodoxy judges with acumen and accuracy when the issue is that of detecting the fragility of others.

Now we shall see something of the vicissitudes of four important figures in rational criticism of relativity: these are the four authors I have chosen to read thoroughly, but other authoritative critics can be traced especially in the early days of the fortune of relativity. One of the critics of the past was Lenard, of whom we spoke in section 5.6, and who threw away his scientific authority by seriously compromising with Nazism. Another was Ernst Gehrcke, not personally compromised but involved in the politicization of the question about relativity that took place in Germany around 1920 and later. Another was Michelson, who was always misunderstood for his obvious conservative predilection for the ether theory: given the conceptual confusion that is aroused by the academic consensus to relativity, it is not strange that many are incapable of distinguishing the issue of the ether and the still unknown process of propagation of electromagnetic waves from the issue of the inconsistency of the attempt of Einstein's theory.

Louis Essen

The four rationalist critics whose ideas we shall now describe with some detail ultimately all say the same thing: they remark that the theory of relativity involves the contradiction we know, and make it useless for any purpose. However, each of them has some particular feature to teach us. The first one we shall deal with is Louis Essen (1908-1997), who deserves a place of pre-eminence given the immense authority that his professional curriculum gives him. In the 1930s Essen, who for his entire career was an official researcher at the British National Physical Laboratory, built the best quartz clocks that could be constructed at that time. During World War II, he was commissioned to test certain radar components by means of a suitable microwave equipment, and this experience brought him to conceive of the idea of a new method for measuring the speed of light: Essen had to fight to see his proposal accepted, but then his result was accepted as the new standard and remained for a long time. Finally, towards 1950 Essen began to deal with atomic clocks, built the first working models, managed to impose his views on the international commissions for defining the standard of Universal

Time, and in the 1960s was considered one of the most authoritative personalities in the world, if not the most authoritative, for everything concerning the technology of time measurements and the construction of clocks[94]. Given this curriculum, it is difficult to imagine that this designer and builder of atomic clocks could not understand Einstein's papers about electromagnetic wave speed and clock synchronization; if Essen were to point out the fundamental contradiction of relativity, it would be very difficult not to give him credit. And this is exactly what happens: in 1971 Essen published a small volume[95] of criticism of relativity in which he summed up the objections he had developed about fifteen years earlier and in which we find the objections we know in addition to various more detailed scrutinies that reveal the superior Essen's experimenting skill. The text is only 24 pages long, but inside there are all the reasons for rational criticism of relativity, which are lucidly identified and expressed concisely, as well as many other reflection items that would deserve to be weighted in further research on the scientific side of the problem of relativity. The 1971 paper is a valuable text for the author's superior competence, but equally interesting is the sketch of human reality that emerges from the account that Essen left of his affair with regard to relativity. In old age Essen wrote a small scientific autobiography that was published[96] by a descendant, and its sixth chapter "Relativity" offers the reader the consolation of feeling in the company of a mind thinking with the simplicity and honesty of true intelligence. In his 1971 essay Essen presented things in this way[97]:

> A common reaction of experimental physicists to the theory is that although they do not understand it themselves it is so widely accepted that it must be correct. I must confess that until recent years this was my own attitude. I was, however, rather more than usually interested in the subject from a practical point of view, having repeated, with microwaves instead of optical waves (Essen 1955), the celebrated Michelson—Morley experiment, which was the starting point of the

---

[94] All this data can be verified in Encyclopaedia.

[95] Essen Louis 1971.

[96] Essen Ray 2015.

[97] Essen Louis 1971, p. 2.

theory. Then with the introduction of atomic clocks, and the enormous increase in the accuracy of time measurements that they made possible, the relativity effects became of practical significance. Corrections are already being made to the frequencies corresponding to the atomic transitions employed in the clocks. The correction for the effect known as time dilatation is about four parts in 10^{11} for the hydrogen maser and four parts in 10^{13} for the caesium-beam standard. The suggestion has also been made that relativity effects should be considered in the definition of the unit of time. Thus there is an important practical reason for subjecting the prediction of these effects to a critical examination.

In autobiographical notes Essen tells us that at university he had been taught something of relativity with the warning that he and the other students would not understand it, and adds that he remained in ignorance as everyone until he felt the need to comprehend the problem for the reasons we saw. Unsatisfied with literature, he turned to Einstein's text, which of course is ED1905[98]:

> ... I, therefore, studied Einstein's famous paper, often regarded as one of the most important contributions in the history of science. Imagine my surprise when I found that it was in some respects one of the worst papers I had ever read. The terminology and style were unscientific and ambiguous; one of his assumptions is given on different pages in two contradictory forms, some of his statements were open to different interpretations and the worst fault in my view, was the use of thought-experiments. This practice is contrary to the scientific method which is based on conclusions drawn from the results of actual experiments.

Essen's first reaction was to think that the theory should be solid in spite of the bad style of the original text, and therefore Essen continued to read widely, until he was convinced that the theory was made useless by internal contradictions and developed a critique similar to what we have expressed in this book. Then Essen expressed his reservations in writing, and thought naively (he says) that scientists would be happy to solve the many confusions that had long persisted, finding out soon that no one paid attention or answered his criticisms, and that all continued to repeat the mental experiments and the logical errors they had been taught by Einstein. More than one hinted that if Essen had gone down this route his reputation and career prospects would likely suffer, so he continued

---

[98] Essen Ray 2015, pp. 63 and following.

searching silently until he retired. In the meantime, he went on to realize that "scientists can be irrational like anyone else," and that by accepting the theory without understanding it they closed their minds to discussion: the understatement is very stylish, and the same concept would deserve to be expressed in a much more efficacious way. Once Essen was invited to give a lecture within the Royal Institution, and on that occasion he received an enthusiastic response from the public and many congratulatory letters, but they were written privately and without use of the letterhead of the organizations to which the correspondents belonged: which Essen says he noticed with some amusement. Moreover, the tone of Essen's writing is pleasant for the good-tempered attitude with which he contains himself by mentioning the less noble features in relativistic literature[99]:

> It is a subject about which writers tend to use more emotive. language than is usual in scientific texts. For example, L.B. Loeb and A.S. Adams (1933) state that most of those who attack relativity are either fanatics or so poorly equipped mathematically that they are incapable of understanding or following the processes involved. The reference here to mathematical ability is puzzling because, as we shall see, parts of Einstein's papers that are often criticized involve no mathematics.

What is to be noted in particular in Essen's vicissitude is that a person with a technological curriculum such as his, always preserves his sound judgment: Essen never passes through the acceptance of the theory. First, he just did not know it and admitted he did not understand it, believing in the prevailing opinion of the scientific world as everyone in good faith does, then he studied accurately the theory, but the study immediately led him to identify the problem without passing through a period of relativistic militancy. We shall briefly see that things are not always so simple, and that the rational criticism of the theory can be less linear and made weaker by misunderstanding. In general, there is a clue that technologists and engineers are the most disposed to realize the absurdity of relativity, while theorists tend to deceive themselves with the aptitude for abstraction and for misunderstanding the meaning of abstract

---

[99] Essen 1971, p. 1.

thinking, which is generalization and criticism of experience, not invention.

Quirino Majorana

Quirino Majorana (1871-1957), whose research and experimenting career extended over many decades and with many interests, shares with Essen the insight of vision regarding the inconsistency of relativity. The designer of prototypes for the electric transmission of images already in 1894, after Marconi's first Transatlantic Transmission of 1901, Majorana was the first to carry out successful radio-telephone experiments, and all his scientific career was at the border between experimental research and technology. He recalls that he was cognizant of relativity already in 1916, and that he designed some experiments to verify whether the light behaviour was ballistic, that is, if the propagation speed of the light is with respect to the source. These experiments did not succeed, but in 1917 and 1918 Majorana published with thorough scientific honesty the accounts[100] of his failure, but did not conclude that Einstein's contradictory interpretation could constitute a theory of the phenomenon. Indeed, the fact that we cannot come to a definite conclusion about the fundamental problem of propagation of electromagnetic waves can be used as an argument for unusual and arduous theories as much as is desired, but not for a contradictory theory such as that of Einstein. We know that the constancy of the speed of light is the expected result with rational motives within the framework of the Lorentz's hypothesis and we cannot rule out that another model compatible with this result be formulated, but we cannot admit that the solution is in a theory for which a phenomenon

---

[100] *Sul secondo postulato della teoria della relatività*, in *Atti della R. Acc. dei Lincei. Rendiconti*, cl. di scienze fis., mat. e naturali, s. 5, XXVI [1917] pp. 118-122, and in *Philosophical Magazine*, s. 6, XXXV [1918], pp. 163- 174; *Dimostrazione sperimentale della costanza della luce emessa da una sorgente mobile*, in *Atti della R. Acc. dei Lincei. Rendiconti*, cl. di scienze fis., mat. e naturali, s. 5, XXVII [1918], pp. 402-406, anche in *Philosophical Magazine*, s. 6, XXXVII [1919], pp. 145-150.

is apparent and symmetrical **and** is real and not symmetrical at the same time. Majorana, though not very optimistic, thought that this concept could be made clear to the scientific community, and instead he too clashed with the deafness of all: with the certainty of the theorists and with the "discomfort of experimental physicists who do not dare often discuss the great mathematical developments of the theory and end up in an intentional personal and apparent agnosticism."[101]

In some lectures held in his old age, in which Majorana expresses himself with the gentlemen's composure of past times, he testifies how astonishment never ends when one ascertains that such a simple and linear reasoning at the reach of everyone as it is which criticizes relativity, is not developed by those who would have the greatest ability to do it[102]:

> But, returning to the simply kinematic part of the theory, it is possible to see certain undoubted contradictions to which it gives rise. This has been the subject of my careful examination for several years, and on this point I would like to draw your attention, while I note from now that these are simple and clear arguments, that with real surprise do not present themselves to the mind of relativity advocates.

The following arguments are the same as we developed, the usual of all rational critics, with some more specialized observation and some more definite opinion that Majorana could afford after a lifetime spent in research; and the "surprise" of the fact that such observations are not shared universally corresponds to the state of mind that will always accompany anyone who frames the situation while remaining faithful to the principle of non-contradiction.

The stories of Essen and Majorana teach us that experimenters and physicists involved with technology find it much simpler and more natural than theorists to control the logical consistency of theories: and this is not an a priori obvious fact, because at first glance it could be thought that, on the contrary, empiricists could be more subject than theoreticians to the temptation of pragmatic compromises with sacrifice of logical cleanness. Instead, this experience shows the

---

[101] Majorana 1956.

[102] Majorana 1951.

opposite: if theorists become fascinated with some theory that appears to be of superior quality to those who examine it superficially, in order to preserve the theory they are inclined to fall into a state of cognitive dissonance and to implement self-deception strategies to avoid discomfort caused by dissonance. This state of affairs depicts the present state of theoretical physics, which nowadays lives in its world of perfectly Baroque fiction, and will remain until a young generation revolts in the name of a neoclassical and rationalistic vision of science. The attempt to force their colleagues to think rationally made by people like Essen and Majorana is inevitable, but is destined for failure, because cultural paradigm changes occur when the energies of new generations want them to occur and fight to produce them: the old never abandon their customary mistakes, and this story teaches in particular that they do not abandon them even in a case where everyone would have the chance to see the enormous and gross contradictions of a theory born from a gimmick conceived by an immature mind as is Einstein's relativity.

Henry Bergson

Henry Bergson (1859-1941) unwittingly found himself in a personal conflict with Einstein and his story witnesses a misunderstanding that is quite common, but in which technical minds such as Essen and Majorana would never fall into. The whole story is interesting, but Bergson's mistake is the most instructive part because it corresponds to a key to acceptance of relativity by theoretical but non-technical specialists.

Bergson's philosophy had great fame in the first twenty years of the twentieth century. It is built on the distinction between the objectified and measured time of science and techniques, and the time lived by the reality of human consciousness. According to Bergson, time measured not only by modern physics, but by any technically rational behaviour, even primitive, is represented as a straight line, and represents nothing less than a fourth geometric dimension: in the objective representation all things are placed in a system having three spatial dimensions and everything moves along the straight line of the fourth dimension, the line of time. Already in his first work of 1889, *Essai sur les données immédiates de la conscience* (in English generally known with the title *Time and Free Will*), Bergson conceived the complex of space and time as a continuous four-dimensional one, without however interpreting it otherwise than according to traditional Euclidean geometry. The time Bergson calls "lived" is different: it is inseparable from memory, and in it things accumulate as records that constantly return to consciousness and continually return to immerse themselves in the unconscious. The difference with the time of object construction is that in the time of memory the past, the present, and the expectations of the future are altogether present, because what is represented by consciousness can only be actual now in the present, while in the time of measure the past, the present, and the future are rigorously external to each other: what is past is in time coordinates different from what is present and therefore is not. Thus the time of object construction is a kind of space: the past, the present, and the future are external to each other in the linear dimension of time as things are in space. Instead, where there are consciousness and memory, past, present, and future expectations are actual altogether. From this way of considering, which belongs to everyone's experience, and which is not at all controversial from a psychological point of view, Bergson deduced profound philosophical consequences, but it does not interest us now either to deepen or question them. Bergson's philosophy does not introduce any irrationalism in the scientific dimension: the process of technique and science through geometric time is not in question; only, for Bergson it represents a point of view about things, not necessarily the fundamental point of view, and certainly not the absolute substance of things.

Bergson was present at the Bologna meeting of 1911, listened to Langevin's report and was seized by curiosity and consequently studied meticulously Lorentz's and Einstein's theories of relativity. In 1922, he published an essay titled *Duration and Simultaneity*[103] with which he thought he would do something useful for Einstein, and on 6 April 1922 he met Einstein in Paris, because Einstein had accepted an invitation from the *Société française de Philosophie* (French Philosophy Society) to attend a meeting with several French physicists and philosophy professors. There is an account of this

---

[103] Bergson 1922 and 1923.

meeting[104], and it is easy to understand that Bergson and Einstein had fallen into a rancorous tension. Einstein gave vague answers to the questions that were asked, Bergson being urged told a few words with little enthusiasm, after pointing out that his original intention was just to attend "to listen", and the party was spoiled for the other participants who attended enthusiastically.

The seeming reason for hostility is that Bergson makes long considerations with respect to the problem of perceptional judgment of simultaneity, which is an always open problem and involves circularity. In fact, on the one hand, a judgment that states two events to be simultaneous or subsequent requires a theory of events, because in the absence of it we could never say that phenomena are simultaneous or successive, but only that we have been conscious of them with a state of attention that included them both; but on the other hand to form a theory of phenomena, it is necessary first to determine them temporally, and thus to give confidence to the perceptional judgments in which phenomena appear simultaneously or successively. This is the central motive of Bergson's interest, and in itself it is neither favourable nor unfavourable to relativity: the detailed methodological study of the conditions of simultaneity judgments in itself would just complement relativity, and therefore Bergson believed to make a good service to Einstein detailing the process by which we compare two clocks locally, which as we know is a fundamental element of relativity. But in the 1922 meeting Einstein was controversial and hostile with respect to this subject, and the real reason is that Bergson, as one sees reading his book in its entirety, was perfectly aware that given the hypothesis of the law of inertia we cannot attribute different local time to two systems in relative motion, because we would not have any criteria to decide which system would have dilated time with respect to the other, and so Bergson had realized that we cannot interpret the theory according to the realist criterion. On the other hand, Bergson was convinced that conventional interpretation would be sufficient to obtain the result of Maxwell's equations invariance. We examined this topic in detail in section 3.14, and we saw that in the conventional interpretation the problem of Maxwell's equation invariance is not

---

[104] See AA.VV. and Einstein, "La théorie de la relativité", 1922.

solved except for the form of the equations, certainly not for their physical meaning. As we know, if we were to apply Lorentz's transformations to Maxwell's equations if spatial and temporal dilatation were only conventional or apparent, we would introduce errors in calculations without solving anything, and indeed we would make the problem graver; but we also know that the quantities at stake are not significant from an experimental point of view, so those who do not fully represent the problem from a physical point of view, wondering what would happen if we really applied the convention of keeping the speed of light constant assuming at the same time that its propagation is ballistic, could fall into the misconception that the formal invariance of Maxwell's equations is a sufficient result. This happened to Bergson, and the same that probably happens to many students: learning the mathematical part of the theory, one gets satisfied enough, and does not become aware of the problem of its physical meaning.

Bergson had no doubt that the theory was to be interpreted as conventional (the term he uses is always "factitious"), and nevertheless the invariance of Maxwell's equations was an acquired result whose validity was not to be questioned, so that in the book he mentions this issue vaguely and never develops any consideration in this regard. For example, in the final note of the first chapter we read[105]: "The truth is that the group of transformations discovered by Lorentz generally ensures the invariance of the electromagnetic equations." This is the tone of the whole book: where he mentions the central problem of Maxwell's equations, it seems that Bergson did not have any interest in deepening it and that he did not think anything about it. Thus the formal invariance of the equations was sufficient to attribute undoubted scientific value to the theory, and for this reason he thought it would make sense to open a serious discussion with Einstein to determine in detail the role of perception and intellectual judgment in judgments of simultaneity. But Bergson was also convinced that the conventional interpretation corresponded to Einstein's true view, and that the realist interpretation was a misunderstanding and a picturesque invention by Langevin and others, and that Einstein would appreciate the aid offered to him to

---

[105] Bergson 1923, passim.

restore the rational and non-fairy interpretation of the theory. About Langevin and the realist turning point of 1911 (which we discussed in section 5.5) Bergson once expressed himself in a conversation; this conversation was with an interlocutor who was talking to him about the confusion between signs and things that according to this interlocutor (not particularly original) modern theories would be inclined to. Bergson replied[106]:

> "Einstein walks constantly on the edge of an abyss, from which his physical sense holds him back; but once he fell there (in a speech of Zurich, I believe, where he committed precisely the confusion indicated by you between the sign and the thing signified, between the figure of light and the event which it indicates), and his intemperate disciples, drawing this view towards philosophy, have built on it an extravagant doctrine."

But since the theory, as we know, requires ambiguity between the conventional interpretation that makes it conceivable and the realist interpretation that gives it a physical meaning, something Bergson did not realize, Bergson's good intentions generated endless controversy and infinite misunderstandings that relativists obviously had no interest in clarifying, of which there is a very detailed account in the essay devoted by Jimena Canales[107] to this affair and its context. It seems, among other things, that Bergson's opinion, which was publicly known, was decisive for the Nobel Committee's deliberation not to attribute the award to Einstein's for relativity, but for the study of the photoelectric effect, a decision which Einstein did not appreciate at all. Canales reports that Bergson was bitter about the affair and did not want to deal with it anymore. Years later, he confided in a conversation that his criticism of relativity had only been understood by "engineers and polytechnic graduates", a very

---

[106] Chevalier 1959, p. 44. "Einstein côtoie sans cesse les abîmes, au bord desquels son sens physique le retient; mais il y est tombé une fois (dans un discours de Zurich, je crois, où il commet précisément la confusion signalée par vous entre le signe et la chose signifiée, entre la figure de lumière et l'événement qu'elle signale), et ses disciples intempérants, tirant cette vue vers la philosophie, ont édifié là-dessus une doctrine extravagante."

[107] Canales 2015.

interesting clue: the testimonies we have do not let us know if any of these technicians have ever succeeded in leading Bergson to understand the uselessness of conventional interpretation. Bergson's words were these[108]:

> "The only people who seemed to understand my book in the scientific world are some polytechnic graduates or engineers who master the use of mathematical symbols without being mathematicians themselves. It is curious that relativity seems to be all the less understood by people who know more about science, which proves once more how indispensable philosophy is to the scientist. And why does this happen? doubtless because our scientists, by their very science, are incapable of understanding *the gap* which exists between the sign and the thing meant."

This diagnosis would need to be deepened: the reason why scientists show no critical sense about relativity is perhaps not a methodological deficit, which would be really too huge (we would attribute to physicists the inability to understand the distinction between an edible roast beef and a painted one, which is a bit too much), but it is in the fascination that relativity exerts for non-scientific reasons and that can generate the will to assert relativity even in the face of the manifest impossibility of its assumptions. Bergson here does not seem to be equipped enough to be aware of the symbolic extra-scientific meaning of relativity, and this is curious because, ultimately, the cultural value of relativity and of its images of time is precisely that of being a surrogate of philosophy for the use of people who cannot read and appreciate either Bergson's philosophy or Proust's novels, for example. However, this detail witnessed by Bergson confirms the suspicion that the intricacies of

---

[108] Chevalier 1959, p. 50. "Les seuls qui aient paru comprendre mon livre dans le monde scientifique sont quelques polytechniciens ou ingénieurs ayant l'usage des symboles mathématiques sans être eux-mêmes mathématiciens. Il est curieux que l'on paraît d'autant moins comprendre la relativité que l'on sait plus de science, ce qui prouve une fois de plus combien est indispensable au savant la formation philosophique. Pourquoi cela? Sans doute parce que nos savants, de par leur science même, sont incapables de comprendre *l'écart* qui existe entre le signe et la chose signifiée."

relativity to be disentangled require the mental form of those who have both a basic scientific preparation and an aptitude for the concreteness of judgment in application problems. The statistical sample is small but unambiguous: it seems that experience with technology is associated with a stronger perception of incoherence, while the theoretical inclination is correlated with the opposite. Moreover, technological experience is correlated with the absence of complacency for a contradictory theory, while the theoretical inclination receives a reward from complexity in itself, and therefore enters into a state of consciousness that removes by self-deception the awareness of contradiction. Technicians, even though they had never thought of it, know that contradiction is neither in things nor in thoughts, but only in expression. If the time dilatation theory had to be applied with respect to relative speed, if the theory indicated significant coefficients for technical practice, technicians would ask: and now, among infinite relatively moving systems, to which of them should time dilatation coefficients be applied? And with respect to which other systems? On the contrary, the problem seems to be too complex for theorists strong in maths, but who have not long cultivated the habit of designing projects and controlling results; in a similar way Bergson's sharp theoretic mind soon became aware of the contradiction of the realist interpretation of theory, but did not notice the uselessness of conventional or apparent interpretation. Yet Bergson shows that he knows exactly what he is talking about: his preliminary description of the theory in Lorentz's version, which Bergson called "half-relativity", is masterful, and in this book we took it as a model (simplifying it slightly) because it is among the clearest and most elegant things that can be read on the subject. Bergson realized perfectly the logical consistency of Lorentz's hypothesis and highlighted with particular elegance that the very generalization of the law of inertia necessitates the abandonment of the concept of local times, which in the context of Lorentz's hypothesis instead is justified and admissible. Finally, Bergson teaches us that studying the version of Einstein without knowing that of Lorentz leads to misunderstanding of what is being read. In this regard, I report a personal observation: as a result of the fact that the relativistic literature mentions Lorentz's hypothesis as a mere curiosity, an expert teacher in relativity and irremovable in the dissonant trust in the theory, when asked revealed that, considering only the M+M experiment in itself, he never thought nor understood that by Lorentz's theory the constancy of c in all measurements of the speed of light is the logical and expected result, and that therefore Einstein did not enunciate something sounding strange to the specialists of his time. These details help to better understand the mechanism by which inaccuracy and bad expression become the culture broth of suggestion.

One last observation concerns the experience I made by reading Canales' book on the conflict between Einstein and Bergson: the book is very informative and is a huge source of stories, analysis, and accurate and very useful data. Starting to read it, it seemed obvious to me that the author would adhere to the rational criticism of relativity, which is the same for all those who succeed in abandoning the favourable prejudice for Einstein, and consists of realizing the contradiction between the apparent and realist interpretation, which can be expressed in very few words by those who have clear ideas. With astonishment, going on I realized that this is not the case: the author is irremovable in relativistic militancy, and writes with so much emotional participation that writing ninety years after the events the readers perceive her feelings of solidarity with Einstein and her dislike of the figure of Bergson. These psychological issues will have to be developed beyond the anecdote when we shall begin to write the cultural history of the anthropological phenomenon of the specialists' consensus on relativity, which is not the bloodiest misfortune, but is one of the strangest and most extraordinary things that have happened to the human species.

Herbert Dingle

The case of Herbert Dingle (1890-1978) dates back to the 1960s, but is still well known for the discomfort that Dingle managed to cause to the English scientific community. Dingle was an astronomer, philosopher of science, president of the Royal Astronomical Society, and professor at University College London; in short, he was a full member of the academic establishment, and he was not a technologist but a theorist. In 1922 Dingle wrote a very simple popular booklet on relativity, and in 1940 he wrote a brief but specialized account. In the 1940 book, Dingle developed the same misconception as Bergson: he built his argument in a conventionalist way, assuming that the realist interpretation was not in question and that no clock really lags depending on relative speed, but believing that the conventional interpretation was useful for something. For this reason the book presents itself differently than usual. After two introductory chapters, the third chapter "Relativity and Length" comes to the point with these words that exemplify the different tone with respect to the usual relativistic literature[109]:

> There must have been some defect in our reasoning: what was it? The answer is very simple: it is that we have used a false definition of length. The quantity that is physically important is not l, as ordinarily defined, but $l\sqrt{(1-v^2/c^2)}$, where v is the velocity of the object concerned in the direction in which the length is measured, with respect to whatever standard of rest we choose to adopt.

According to this way of exposing the theory, the constancy of c is accepted being the experimental data of the measurements made for any beam with respect to whose source the observer is at rest, and to extend the constancy of c to the case of motion with respect to the source of the beam, the ancient convention with which a unique length measure l was reached was replaced by a new convention for which the length is always transformed according to Lorentz, and so on. This idea would not lead to consistent results, because in any case in the dual hypothesis of not introducing the ether and accepting the constancy of c the application of corrections through Lorentz would only make sense for measurements performed through electromagnetic waves, while for any other measure we should continually correct our samples of space and time; but the book does not concern itself of this, and in the following parts it does little more than present the form of Lorentz's transformations and their mathematical consequences, so there is not much to judge which results Dingle could reach given this starting point, that made unstable every measure making it depend on any speed among the infinite possible speeds. But in general the book is simply insensitive to the question of the physical meaning of the mathematical forms it illustrates.

---

[109] Dingle 1961, p. 23.

7. Success of special relativity and opinions

Then, after ten or fifteen years, Dingle noticed that his colleagues took the realist interpretation for granted, and then he finally realized the state of affairs. Not only did he notice his colleagues' illusion, but also that the conventional or apparent interpretation would make the theory completely useless. Years later he would write this way, abandoning the conventionalist interpretation[110]:

> The supposition that the theory merely requires each clock to *appear* to work more slowly from the point of view of the other is ruled out not only by its many applications and by the fact that the theory would then be useless in practice, but also by Einstein's own examples.

The book of 1940 has several curiosities. It was published not by any publisher, but by Methuen, who published books by Einstein and relativists in the United Kingdom: so we can see that nobody realized that the conventionalist interpretation that pervaded it was heterodox. In fact, later Dingle remembered that his 1940 book had received the compliments of Max Born, the ultra-orthodox friend of Einstein, and wrote[111]:

> When, in 1940, I published my second book on the subject (*The Special Theory of Relativity* — Methuen), now in its fourth edition and still much used in universities in this country and in America, Max Born wrote me: 'I have enjoyed it very much, as your explanations of the difficult subject are very clear and well presented. I hope the book will find many readers.'

In fact, despite Dingle's rethinking, the publisher who probably had the copyright reprinted the book again in 1961 (sixth reprint), allowing Dingle to add a seven-page preface in which the author disowned his work of twenty years before[112]:

> Since this book was written, reasons have appeared, which to me are conclusive, for believing that the theory is no longer tenable. Though this is not yet generally accepted...

The situation is more unique than rare, and alone is a good witness to the general confusion that lies within relativity: if Dingle had become a radical opponent of theory, how could the book still have a

---

[110] Dingle 1972, pp. 45-46.

[111] Dingle 1972, p. 106.

[112] Dingle 1961, pp. v-vi.

market? Did the publisher not have available one of the infinite orthodox titles that are continuously written? And why was a book that had been arguing in a heterodox way and that the author had disowned still used as a textbook, as Dingle told us? The behaviour of the publisher and of teachers who used the book recalls the sects of followers of the science fiction writer Howard Lovecraft: these zealous adepts wanted to believe that the writer was a prophet of revelations of true supernatural things, when he always insisted that the monstrosities described in his books were purely fanciful.

The book's preface in the 1961 reprint is designed to save Dingle's honour and at the same time justify with some reason the book reprint, and the result is inevitably hypocritical. First of all, the reason why relativity is declared no longer sustainable is attributed to the fact that recent experimental data would support Ritz's theory and the constancy of c with respect to the emitting body: which is a pitiful lie because in-depth experiments to verify whether Ritz's ballistic theory is founded have never been made, and serves to avoid talking about the intrinsic contradiction of relativity that Dingle had realized in the meantime, but could not be mentioned in the preface of a book that still served to teach relativity. Meanwhile, Dingle had also noticed the intimate absurdity of the conventionalist interpretation itself, as he wrote this way:

> The special theory of relativity as here presented is based on what was believed to be the discovery that the important quantity is not L but $L\sqrt{(1-V^2/c^2)}$, where V is the velocity of the body. Special relativity thus appears as a perfectly normal department of physical science, extending and making more exact the system of relationships between measurements that constitutes the body of physics. There is, however, one difference between velocity and the other quantities, such as temperature — namely, that you can make it what you like. You have no choice with temperature, as we have defined it: your thermometer will tell you without ambiguity what it is. But velocity has no meaning apart from an accepted standard of rest, and the principle of relativity is the principle that there is no such standard fixed by nature but that you may adopt any standard you wish. That makes the 'length' of a body indefinite, and that means that all other physical measurements that are definitely related to length (i.e. all other physical measurements) must share that indefiniteness. The main bulk of this book is occupied with the determination of the indefinite expressions (i.e. expressions which

are functions of velocity) for the other common physical measurements, such as those of time, mass, etc.

Speed cannot be taken as a fundamental quantity in the conventions about measurements, regardless of whether or not it is an absolute constant. Dingle here attributes this change to the abandonment of the hypothesis that c is an absolute constant, but this is not a relevant topic because the conventional interpretation would make the length measurements unstable in any case. The truth was that having meditated on relativity, Dingle had also realized the stratagems he himself had invented to make it acceptable, which in his case were unusual because they were designed to make the conventional interpretation plausible, but in any case they were gimmicks that served to support relativity, perceived as a value in itself and not as a tool to find solutions to scientific problems. Essen may have influenced Dingle in this regard, because Essen, who was an expert in measures and of whom Dingle had a very favourable opinion, had focussed part of his criticism on the specific issue of conventions and the nonsense of taking a speed as a conventional unit[113].

Beginning in 1958, Dingle literally began to torment relativist colleagues in UK universities by expressing the issue in the simplest way of all: given two systems A and B in relative motion, do you want to tell me how we know which of the two clocks will delay as a function of speed, assuming no one will argue that clock A lags over clock B **and** clock B lags over clock A? These are Dingle's words[114]:

> According to the theory, if you have two exactly similar clocks, A and B, and one is moving with respect to the other, they must work at different rates (...), i.e. one works more slowly than the other. But the theory also requires that you cannot distinguish which clock is the 'moving' one; it is equally true to say that A rests while B moves and that B rests while A moves. The question therefore arises: how does one determine, consistently with the theory, which clock works the more slowly? Unless this question is answerable, the theory unavoidably

---

[113] Essen's paper was published in 1971, but Dingle is likely to have been aware of the criticisms that his then-famous fellow and compatriot had been expressing since the 1950s onwards. See Dingle 1972, pp. 114-115.

[114] Dingle 1972, p. 17.

> requires that A works more slowly than B and B more slowly than A —
> which it requires no super-intelligence to see is impossible.

As expected by Dingle, no one dared to argue that both clocks delayed, but contrary to Dingle's predictions no one ever gave a clear answer to the question: but since Dingle could not be ignored given his past and his academic career, it began a controversy made of diversions and irrelevancies in which those who were urged by Dingle tried to prove that Dingle did not understand the theory. Max Born and professor McCrea publicly replied to Dingle with the argument: it is pointless to ask which clock will lag between the two, because the problem is not this, but is to study and understand Minkowski's geometry[115]. That is, the solution would be to consider acceleration as the determining reason for a phenomenon that must, however, be described as a function of speed, with all the science-fiction consequences that this entails, and that we have seen above in section 6.4. Dingle could rightly say that his elementary question never received anything that resembled a pertinent answer.

Dingle's other prediction that never happened was that anyone was urged by his writings to reflect on things and to realize the absurdity of the theory, or to change their mind. Nobody, no teacher and no student changed their mind, and the controversy continued until the English scientific press stopped publishing Dingle's articles. The failure to arouse abandonment of relativity was total, and Dingle wrote this way[116]:

> Being a poor psychologist, I did not realise that scientists, like other people, are far more ready to search for flaws in other people's reasoning than to eliminate prejudices from their own, and I remained in a state of bewilderment at my inability to make clear to others what seemed so obvious to me.

Right, but there is much more: the human pride that is encountered in ordinary cases is not enough to explain this extraordinary story. People never change an idea without having an incentive to do it and some pretext to attribute not to themselves, but to the world the responsibility of their past mistakes. People who change their minds

---

[115] Dingle 1972, Appendices.

[116] Dingle 1972, p. 41.

must receive a reward for self-esteem or otherwise, which in the case of consent to relativity cannot exist unless the return to reason does not come from a personal development towards maturity and this process is in itself a reason for pride. Let us put ourselves in the shoes of people who know a lot about math and physics (theoretical) and who to follow the prevalent social tendency of their environment and of the whole world have generally fallen into the state of cognitive dissonance that allows them to pronounce and write the repertoire of contradictions of relativity: why should they humble themselves by acknowledging their mistake just because another tells them? The situation is not that which happens when things have changed outside us, and therefore a theory or an opinion can be repudiated because some objective external circumstances force us to change our interpretation. When this is the case, those who change ideas can tell themselves and others that they are keeping up with the world, and that the opinion change is a sign of intelligence and an open mind. But relativity, like mustard gas, is a poisoned fruit of the systemic cognitive disaster that took place in the years of the Great War, and the experts subjectively adhere to it only because they do not have sufficient independence of judgment to recognize its absurdity against the prevailing opinion. Therefore, relativity is rejected only by those who become aware of its absurdity with their own strengths, and it will only fade away when unprejudiced young people refuse the decadent taste of today's academy, and will be the leaders of a reversal of the prevailing orientation. All this is far in the future, because as we know who has the endowments of intelligence and judgment independence that are necessary for *not* understanding relativity usually leaves it out of study, given the strength of rhetoric that was developed to create an appearance of rationality and depth to this theory; Dingle, a poor psychologist, did not realize that he should try to speak not to those who believe having understood relativity, but to those who realize they are unable to find a coherent theory and at least know they are perplexed.

In 1972, *Science at the Crossroads* was published, the melancholic and gloomy book that Dingle had been[117] "trying for more than thirteen years to avoid having to write", that tells the story of the

---

[117] Dingle 1972, p. 15.

controversy with the British scientific world, and develops the criticism of relativity with great detail. The first thing we observe in the book is that in the midst of other themes we find this incidental observation, placed in brackets. The time referred to is that of the explosion of Einstein's fame after 1919[118]:

> ... I myself was for long at fault for failing to recognise that the mathematics of the time was simply another form of mediaeval logic restored to its old position of authority over experience. I offer no excuse for this, nor do I regard my own intellectual pilgrimage as of sufficient general importance to relate here the course it took and how later I awoke from my dogmatic slumber.

We see that Dingle acknowledges that he was wrong as many were, and he recognizes that he has no justification, but he really proves to be a poor psychologist, as he admitted elsewhere, because the description of his personal intellectual pilgrimage would be very interesting to the reader and the reason why he did not want to talk about it is not modesty, but probably it is precisely the inability to find the appropriate words to describe his previous state of consciousness in relation to relativity. It would be very useful to have a detailed description of the process through which Dingle realized that in the past he had accepted an absurdity that another dimension of his inner life knew untenable, and realized that he had created words to support the preconceived opinion he wanted to maintain, without the supervision of the logical component of his intimacy would fully allow it. Bergson says, not about Einstein, but as a general remark[119]:

---

[118] Dingle 1972, p. 96.

[119] Bergson 1889, Chapter II. Translation by F. L. Pogson, London 1913. "Les opinions auxquelles nous tenons le plus sont celles dont nous pourrions le plus malaisément rendre compte, et les raisons mêmes par lesquelles nous les justifions sont rarement celles qui nous ont déterminés à les adopter. En un certain sens, nous les avons adoptées sans raison, car ce qui en fait le prix à nos yeux, c'est que leur nuance répond à la coloration commune de toutes nos autres idées, c'est que nous y avons vu, dès l'abord, quelque chose de nous. Aussi ne prennent-elles pas dans notre esprit la forme banale qu'elles revêtiront dès qu'on les en fera sortir pour les exprimer par des mots;

7. Success of special relativity and opinions

> The beliefs to which we most strongly adhere are those of which we should find it most difficult to give an account, and the reasons by which we justify them are seldom those which have led us to adopt them. In a certain sense we have adopted them without any reason, for what makes them valuable in our eyes is that they match the colour of all our other ideas, and that from the very first we have seen in them something of ourselves. Hence they do not take in our minds that common looking form which they will assume as soon as we try to give expression to them in words; and, although they bear the same name in other minds, they are by no means the same thing.

If Dingle had left us the story of the whole process that had developed within himself, if he had told us how he had woken up from "dogmatic slumber", he would have left us a witness of universal value, because the sensitivity to the fascination of time dilatation is a universal phenomenon, and the renunciation of logical vigilance by skilled and expert people is a universal phenomenon of our time. More important, Dingle's witness would have been especially useful because he had created a variant of the conventional interpretation by which to silence his latent suspicion of the inconsistency of the theory: so Dingle was always warned about dissonance and failed to silence it by repeating the current self-deceiving strategies, but he felt the need to create a personal one. Only, sincerity is not easy for those who should describe a mistake that is painful to them, while for example a person like Essen, who did not have to reproach himself for having ever fallen into the trap, did not even have trouble telling with simplicity how things went: we saw that Essen first considered the theory consistent given the general consensus although he knew he had not understood it, then when he decided to deepen his understanding of the matter because the precision of his time measures demanded this control, he immediately realized that things did not balance out and did what was needed to make the terms of the problem clear.

Science at the Crossroads is divided into two parts: the first concerns the "moral issue", the second the "intellectual" one. The first part has the title of "moral issue" because Dingle's fundamental

---

et bien que, chez d'autres esprits, elles portent le même nom, elles ne sont pas du tout la même chose."

scruple was that in the era of nuclear physics it would be a duty for anybody responsible to recognize a mistake as obvious as what was committed, and to stop to train the new generations of physicists by teaching them to assimilate the contradictions of relativity. An unexceptionable purpose, which turned out to be utopian. But let us read this passage[120]:

> Another generalisation, which I think has an important significance not only for the question of the truth or falsity of special relativity but also for the matter of the education of scientists, is this: the readiness to respond to my criticism decreases steadily with increasing distinction of those who read it. The leaders in the subject reply, if at all, only when pressed, and as briefly as possible. Those of intermediate status cite experiments of greater or less irrelevance or present calculations of greater or less complication and with no relevance at all. Students and young Ph.D.s are vociferous. 'After your argument in *Nature* with Professor Max Born' (...), wrote the former editor of *Nature* to me in 1963, 'I had a large number of communications and quite a number of individual unannounced visitors at *Nature* office. As you implied in your letter they each one felt that he could prove you were wrong in your view and each one got about it in a different way... all the people who submitted communications or wished to discuss this problem with me, could scarcely be considered first class men of science as compared with Max Born.'
>
> I think this shows how, even in science, what at its beginnings is recognised as a speculation, with greater or less plausibility, develops with time into a compulsory dogma, which whosoever disbelieves thereby brands himself as an ignorant fool.

The presumptuous embarrassment, disguised as lack of interest, of academics is not surprising at all. The scruples of students and young scholars, however, are overwhelming in their desire to defend the theory, which certainly happens for the generic reason indicated by Dingle, but which in this case probably is also related to the protective character of Einstein's myth after 1919. Young people did not seize the opportunity to express their potentiality through the controversy with the generation of their fathers that was offered by Dingle, but on the contrary they hurried to repair the sin of disbelief committed by Dingle and to protect themselves from the temptation

---

[120] Dingle 1972, pp. 38-39.

to follow him. They took refuge in their shell under the holy image of Einstein, because those who internalized the state of self-protection from dissonance through the illusion of understanding relativity, believe they are mastering the most complex thing that exists and hence receive a self-definition that they cannot renounce to, and to a greater extent this happens in the context of the fragility of all values typical of the society of our time. The lack of self confidence of young people and the fragility of their skills do not allow the luxury of independent judgment in a matter where the consequence of independence is the return to ignorance, albeit Socratic, and especially to solitude. After having long spoken of the attitude of the most authoritative academics, Dingle adds[121]:

> I have omitted all reference (...) to the prevailing frame of mind of students and young research workers in this subject of which I have had ample experience over the years through visits to universities to address student societies and discuss the problem with them. It will suffice to say that this is, in one respect, the most saddening aspect of the whole matter, for it is evident that the students have been trained, consciously or unconsciously, to believe that criticism of special relativity is a sure sign of ignorance, not to say stupidity, on the part of the critic (...); only comparatively rarely have I been asked questions for information.

This first part of the book bitterly describes Dingle's controversy with the English scientific community, but the narrative is never spoiled by the pettiness and petulance typical of polemic and controversial writings, and indeed captures the readers' interest and is a source of a lot of data for those who want to begin to understand the psychological and anthropological aspect of the phenomenon: in this regard, Dingle's account is only a witness and does not contain nor pretend to contain any attempt of a theory, but this is more of a merit than a fault, since every attempt to create a theory would have been premature. It will be up to us tomorrow to form a theory of the specific cognitive dissonance affecting the scientific elites. One of the most impressive witnesses described in the book is Kathleen Lonsdale, an outstanding Crystallography specialist still remembered (see Encyclopaedia), to which Dingle could personally address as a colleague; let us remember that Dingle was a senior member of the

---

[121] Dingle 1972 p. 114.

British scientific community and had a privileged point of view on the problem[122]:

> I then approached Kathleen Lonsdale, who was not only a former colleague at University College London but also a close personal friend. Her work also was only indirectly related to relativity, and she shared the general belief in its essentially mysterious essence in even greater measure than most, for it had been presented to her in her student days cloaked in such metaphysical irrelevancies that, being naturally predisposed to ascribe the appearance of nonsense in the instruction of her tutors to her own incompetence rather than to actual fact, she had been rendered unable to hear the word 'relativity' and retain her power of simple reasoning. 'It interested me so little that I forgot it as quickly as possible,' she wrote. Nevertheless, she agreed to read my paper and try to follow it, but found herself powerless to do so. 'My difficulty is that I get so far and my mind goes blank. I never would have supposed that it was so difficult to read and understand something outside one's own field.' 'I spent about six months trying to make sense to myself of your paper, but each time I tried, my mind just went *blank*. Apart from trivialities I could neither criticise nor approve it. The best that I could say to myself, to justify my communicating it at all, was that — for what my judgment was worth — I could not see the fallacies in it, if there were any... If I were to spend six weeks reading it again it would still mean nothing to me. My mind is not built that way. The whole of Einstein's theory just seems esoteric nonsense, as far as I am concerned... My mind simply does not *care* whether clocks go at the same rate or whether they don't, and it refuses to work when I try to make sense of it. I'm sorry.'

Let us now recall that the question with which Dingle troubled interlocutors simplified the question in the most elementary way and simply asked: given the premise, the principle of relativity and symmetry, how can we decide which will lag between a clock A and a clock B both moving one with respect to the other?

> Kathleen Lonsdale was one of the most intellectually honest people I have known, and that her mental endowments were too slight to enable her to follow the simple piece of reasoning given on p. 45 is too ludicrous to be entertained by anyone who knew her or even knew of her. The fact that she could write in these terms is an outstanding testimony to the harm which has been done by the illusions that are so

---

[122] Dingle 1972, pp. 56-57.

widespread concerning relativity. We shall see that it is general; the more distinguished and the more mentally honest and the more concerned in their work with the special theory of relativity the experimental physical scientists may be, the more convinced they are that the theory is unintelligible to them. What they cannot transcend is the conviction that the 'mathematicians' do understand it and cannot be wrong: they choose to believe themselves fools rather than that. We shall meet with other examples; Kathleen Lonsdale merely expresses it more starkly.

In front of this testimony we must pay attention to the state of affliction that the matter provoked in Lonsdale. We read between the lines that a person like this has the intelligence resources necessary to withstand rhetoric that disguises cognitive dissonance, but does not have the strength to accept the inevitable consequence: the disillusion about many people who appeared authoritative, and the fall of some cultural references and some certainties considered fundamental. Dissonance is a very unpleasant state of consciousness, but the price to be paid to become substantially free from it can be so high that it is preferable to silence it even through self-esteem sacrifice: as Dingle observes, the most competent prefer to believe they do not understand the theory instead of judging it, with a rare reversal of what is usually observed.

The second part of *Science at the Crossroads* as can be expected, is centred on the elementary contradiction we know, but discusses the problem of relativity by entering into many interesting details with competence and thorough knowledge of the history of science. In general, the whole book contains intelligent reflections on the abandonment of the experimental method by the theoretical physics of the twentieth century, the progressive separation of theoretical physics from experimental science and technology that proceed according to classical methods and deep misunderstanding of the function of mathematics with respect to the knowledge of the physical world, which is observed in theorists, and whose beginnings Dingle even attributes to Maxwell, who in his opinion gave a sort of idealistic interpretation of the experimental data of the little mathematical Faraday: an idea that suggests an interesting starting point for a research that could be performed[123]. Of course these ideas are in radical contradiction with the epistemological neopythagoreanism of our day, and it is not today but tomorrow that they will be re-evaluated by some generations that will come back to a concrete attitude regarding what actually happens in the empirical object construction processes that are crowned by success.

One of the most interesting themes in the second part of the book is the reconstruction of the reason why relativity received the very first credit opening for reasons intrinsic to Maxwell's equations problem. It should be said that Dingle has no great sensitivity to the extra-scientific symbolic value of the theory, and he never speaks about it. Here too he could have let the witness of a man who was contemporary to events, but in his writings we find only a few hints, such as this one that we read not in *Science at the Crossroads*, but in the introduction written by Dingle for an English edition of Bergson's essay on relativity. To understand the quotation, remember that the two twins of the paradox are often called Peter and Paul[124]:

> ... when the Peter and Paul problem was first posed— by Langevin in 1911— there were circumstances that prevented it from appearing as a serious threat to the relativity theory. To begin with, the possibility that velocities sufficient to cause an appreciable difference in rate of aging would ever be attained was so remote that the problem could not be regarded as other than a *jeu d'esprit*, in quite a different light from that in which we see it now. Hence it called for no more than a token answer, and this was at hand in the circumstance that, in order to return, Paul would have to undergo an acceleration: the theory, as it then stood, was applicable only to uniform motions and so was not menaced by this fanciful case.

> It is easy to say now that the magnitude of the effect was immaterial and that it should have been realized that logically an infinitesimal difference in rate of aging was just as fatal to the relativity postulate as a very large one. It did not so appear then, as I, who remember that time, can testify.

---

[123] Dingle 1972, p. 132.
[124] Dingle 1965, p. xxvi.

7. Success of special relativity and opinions

In addition to being generic, the hint seems not fully exact: in the past one hundred years, the obsession with the promise implied in the symbolism of space-time seems to have been not at all a game, and the same the turning towards the realist interpretation of the theory by Einstein. Instead, this progressive process of mathematization of physics and of misunderstanding the meaning of mathematics is at the centre of Dingle's epistemologist's interest and gives rise to more interesting considerations[125]:

> It is usually taken for granted that the processes of mathematics are identical with the processes of reasoning, whereas they are quite different. The mathematician is more akin to a spider than to a civil engineer, to a chess player than to one endowed with exceptional critical power. The faculty by which a chess expert intuitively sees the possibilities that lie in a particular configuration of pieces on the board is paralleled by that which shows the mathematician the much more general possibilities latent in an array of symbols. He proceeds automatically and faultlessly to bring them to light, but his subsequent correlation of his symbols with facts of experience, which has nothing to do with his special gift, is anything but faultless, and is only too often of the same nature as Lewis Carroll's correlation of his pieces with the Red Knight and the White Queen — with the difference that whereas Dodgson recognised the products of his imagination to be wholly fanciful, the modern mathematician imagines, and persuades others, that he is discovering the secrets of nature.

Given this mental predisposition that was developing, when Einstein "who had no qualms about abolishing the ether and still retaining light waves whose properties were expressed by formulae that were meaningless without it"[126] proposed to eliminate the ether and to maintain equally c as absolute constant, the idea found a favourable ground because what was perceived as the primary purpose was to preserve Maxwell's equations, not the corresponding physical theory[127]:

> As I have said, this theory made no general impression at all at first, so the apparent absurdity of this called forth no appreciable protest (though

---

[125] Dingle 1972, pp. 127-128.

[126] Dingle 1972, pp. 165-166.

[127] Dingle 1972, p. 155.

I remember hearing Sir Oliver Lodge satirising it before Einstein's general theory brought his special theory into prominence), but the fact that it could have been proposed at all is inexplicable until we remember the nature of the acceptance which Maxwell's theory was accorded at that time, which was so well expressed by Hertz — 'Maxwell's theory is Maxwell's system of equations'. The *physical* part of the theory was expendable; only the equations needed to be saved. Einstein saw a way of saving the equations, and did not consider it worth while to 'explain' light.

Consequently, the mathematization of methods led to the whole set of misunderstandings of relativity[128]:

... subjectivity and objectivity were hopelessly mixed up by the conversion of co-ordinate systems into 'observers'. It could not be escaped, because the indications of the Maxwell-Lorentz theory which was universally accepted (except where quantum theory made it necessary to deny them) needed the correction of the Lorentz transformation to make them accord with experimental results, and the Maxwell-Lorentz equations, having been accepted, in accordance with Hertz's perception, as a substitute for a theory, could properly be corrected by another system of equations without too much attention being paid to the absence of any intelligible idea behind them.

The methodological revolution initiated at that time consists of making it legitimate to claim that a mathematical structure has physical meaning without the need to build a model that makes that meaning conceivable in logically satisfying terms. But if we think of it just for a moment, we realize that if we transform the role of mathematics by promoting it from criterion of distinction between possible and impossible events in experience to the criterion of discovery of truth and of truth itself, then, since consistent mathematical constructions are infinite, the criterion of truth becomes and can be nothing more than the social consensus for the words of those who best capture public goodwill. And so we realize that in the glorious anti-classic physics of the twentieth century there are many of the worst features of the era of mass culture.

Describing the intrinsic and methodological reason why relativity might seem plausible to specialists, Dingle passes through an important remark on the character of Einstein's text[129]:

---

[128] Dingle 1972, pp. 181-182.

> ... his weakness (...) lay in his relative inability to follow up the implications of his insight and in a too great readiness to accept a promising starting-point as an achieved goal. He was rather like one of a body of men imprisoned in a dungeon, who alone perceives an opening offering a means of escape, but omits to verify that it does not lead merely to another part of the dungeon.

This observation deserves to be developed beyond Dingle's hint. What we have seen of Einstein's text corresponds to this description: concentrating on the page he is writing and the intuition of something that might be promising, Einstein never controls the coherence of the complexity of the paper or the theory that he is developing. Commenting on ED1905 above in section 3.11, we noted that after describing the phenomenon of dilatation as apparent, Einstein proceeds to use Lorentz's transformations for Maxwell equations without saying a single word to justify the passage: "Wenden wir auf diese Gleichungen die in §3 entwickelte Transformation an...", or "Let us now apply to these equations the transformations developed in §3...", as if it were the most natural thing of the world and its legitimacy were obvious. Yet, we know that the operation is as absurd as it would be for a tailor to see a person getting smaller as he or she walks away and to judge that it will suffice to have ten centimetres of cloth to make a coat for him or her. From a psychological point of view, it can be assumed that this attitude of passing from one assertion to another without being aware of the inconsistency of the whole text corresponds to a singular cognitive disorder that probably afflicted Einstein's personality and lasted all his life long. But what is relevant to the life of the world is not the psychology of any author, but the meaning that is built into the readers' minds given the character of an author's text. Einstein contradicted himself in a direct and immediate manner, without modesty and without any scruple of justifying controversial assertions, probably because he did not realize the problem on account of some serious cognitive impairment: but the effect is that the text, written in good German with long periods and intricate syntax, takes with its own strength a serious tone and an aura of depth and authority. A text that dares to contradict itself brilliantly

---

[129] Dingle 1972, p. 150.

and without looking for arguments intimidates the readers and stimulates them to attribute to themselves the lack of understanding much better than a text that tries to corroborate contradictions with fragile arguments. We have had an example above (sections 5.4 and 6.7) in the argument by Tolman, who had been overwhelmed by the fascination of relativity, but sought justification by humbly asking the reader to accept the unusual character of certain assertions and arguing that black is white to draw physically significant consequences from the set of premises that he himself declared fictitious and conventional.

The bold way in which Einstein claims anything he wants has been observed by the orthodox literature too, which obviously interprets it in the opposite way as a sign of genius. In a widespread book, Pais's, we read this assertion that sounds perfectly complementary to Dingle's remark[130]:

> What is so captivating about the Einstein of 1905 is the apparent ease and the *fraîcheur* with which he introduces new ideas. (...) If the velocity of light does not seem to depend on the velocity of the emitter, then why not to make that into a postulate?

That is right, why not postulate anything that dispenses us from experimentation and criticism of experimental results? Especially when things are such that experimental verification cannot be done? The complacency of this attitude is the expression of an artificial mentality as was the case of the Jesuits that scandalized Pascal's meek Provincial, who had maintained respect for the correctness of logical principles and syllogisms and for the seriousness of thought.

Here we conclude this account of the work of four rational critics of relativity chosen a little by chance: the work to be done to understand how all that we saw was possible is only at the beginning, and the history of the formation of the myth of relativity will require enormous amounts of reading, in which other critical positions will be encountered, probably all converging towards the very simple central arguments. Readers who came here wanting to deepen their understanding might first like to read Essen and Majorana's contributions, which will not offer any difficulty in understanding

---

[130] Pais 1982, § 7a, p. 141.

and in which the reader will see how our arguments can be expressed by others with different personality, but with identical results. Then read Dingle carefully, because this reading (also without any difficulty for those who understood this book) is a first step in the direction of understanding the anthropological side of the phenomenon. Those who want could also read Bergson's elegant and intense prose, which will be less easy, and will probably be useful to reflect on Bergson's philosophy rather than on relativity.

7.5   *Counterintuitive geometry of space-time*

Those who begin the study of relativity on the basis of the knowledge they have received only from the simplest or merely descriptive sources believe that the fundamental problem to be overcome is that concept such as that of dilated or contracted space and time that have become dependent from motion, or Minkowski's correlation between the coordinates of the four dimensions x,y,z,t, are all notions that cannot be represented intuitively and through literal examples. First of all, with reason, people think that the central issue is: can I figure out how time can be dilated? Will I understand how space-time can be described by a counterintuitive geometry? And there are other legitimate questions of this kind, which are all different points of view around the same problem, since everything originates from the postulate of the absolute constancy of c even in the ballistic interpretation of the motion of light.

As we have seen, there is no trace of such problems in the rational criticisms of relativity, and the reason at this point should be immediately clear to the reader: the contradiction between the conventional interpretation and the realist interpretation that the theory is forced to ignore to pretend to have found a kinematic corresponding to the postulate of c constancy in the absence of the ether is so grave that there is no reason to discuss further problems. The theory forces those who want to accept it into a state of consciousness for which they then will claim that time dilatation is apparent *and* is real and that the resulting clock delay is symmetric *and* is asymmetric. Those who resist the fascination leading to this state of consciousness no longer recognize the plausibility of the theory, and therefore they have no reason to discuss any problem about the non-intuitive and non-Euclidean geometry that would describe space-time: if there were a theory that satisfies minimum consistency requirements, then we would have to scrutinise this second problem seriously, but the theory given to us by Einstein gives no elements to reflect on, while Lorentz's theory is completely comprehensible in classical terms and does not give rise to the problem of any geometry contradicting intuition.

Historically however, it must not be overlooked that in the early twentieth century the rational criticism of relativity often used the argument of the counterintuitive character of space-time geometry to reject the theory[131]. Bergson at that time took the point with accuracy, and realized that the law of inertia was incompatible with Einstein's second postulate, but his was an isolated voice and most people who criticized the theory appealed to its contrast with common sense, which is a poor topic, and if applied coherently would deny the value of any scientific abstraction that goes beyond immediate intuition. This is important evidence of the general confusion that was raised by relativity: to avoid resorting to the little clever argument that calls for common sense, it is necessary first to carefully analyse the meaning of the theory and to distinguish its elements presented amalgamated inside its incoherent speech. This rarely happened. But in itself the question of non-Euclidean geometries and their meaning is simply not relevant to us, and we shall only say something to complete this with regard to relativity.

We have all been taught and become accustomed to the idea that there is a three-dimensional space in which we can draw parallel lines that will never meet, that all that exists is contained in this three dimensional space, and that everything that exists is subject to continuous transition from one state to the next one in the linear dimension of the unique time. To tell the truth, initially we use the strange expression "time flows" and often use the river metaphor for time, but as soon as our experience becomes more accurate by performing some measure of time and calculations related to it we come to realize that our attitude was naive: it is not time that is flowing, but things that change state continuously within the dimension of time, which is as firm as space. And then, when we come to this result, we also learn to conceive that the ancient classic representation of Greek philosophy might be an exact description of things in themselves: everything could be immobile in the four dimensions, the three spatial and the temporal one, and the ability to perceive things in their immobile eternity was denied only to the human consciousness, which was given the fate to always follow the sequence of events in time, the ephemeral place of phenomenal

---

[131] A number of variants of rational criticism in the 1920s can be found in the collective volume *Hundert Autoren gegen Einstein*, 1931.

appearance. But let us leave aside this small (semiserious) excursus in the realm of metaphysics. When we think classically, what we experience is that given two parallel lines, the idea that there is a limit beyond which they will cease to be parallel and converge will seem as absurd as the one that there might be a number deprived of the next unit, that is the last number. Let us think of a number having one thousand billions zeros: we know that we can add another unit if we want, and if anyone who states the opposite, we would judge to be lacking in abstraction, not ourselves in error. The same is true with two parallel lines: after extending them in thought for one thousand billion light years while keeping them parallel, nothing prevents us from thinking of adding to them another centimetre if we want, and whoever denies it seems to be more on the road to regress to the mental state of the Boeotians, which according to the Athenians could not count beyond number four, rather than on the road to overcoming the classical vision of the physical world by replacing it with a very modern and unprejudiced one. There is, however, a consolidated misunderstanding since the beginning of the story of non-Euclidean geometries: traditional geometry is understood as if it said that there is a physical thing, called space, which would have the property of being infinite and contain infinite parallel lines that remain so indefinitely. But we do not know if this exists: we only know that the rule that allows us to plot two parallel lines without being able to conceive of a limit to stop them is within our minds and becomes operative as soon as we acquire the ability to distinguish the mathematical form from the physical and perceptional content of experience. Therefore, there is no difficulty in conceiving physical reality as described by any equation that bends all things inside space, or that slows down or accelerates the phenomena of a given system of the same factor so that the system may have a contracted or dilated local time, and all other representations of this kind.

Non-classical physics of the twentieth century, however, offers us a more radical step forward: it makes us suspect that through learning we could no longer believe in our ability to extend indefinitely two parallel lines if we want; it hints that we could become capable of constructing intellectually such concepts of space and time that are no longer congruent with the space and time that we are used to constructing in intuition through the classical view of things. Is all this sensible? Should it be taken literally, or would it always be metaphorical? We are now unable to give general answers to this problem: however, we have come to the conclusion that it makes no sense to address this problem in the face of the theory of relativity, whose frightening coherence deficiency exhausts every speech about it. A theory that used non-intuitive geometries and through these produced logically consistent results verified in experience, would require that the problem of the meaning of non-Euclidean geometries be taken seriously, while inconsistent relativity does not give us any reason to reflect on this methodological problem.

7.6   *Experience of the author of this book*

With all that we have seen, those who briefly analyze the sense of relativity can only conclude that it is a collective illusion that has lasted more than a hundred years and is based on an ambiguous text and on a derivative literature that has consolidated the ambiguity and habit of accepting passively the contradictory background. We know that the theory has responded to some understandable expectations from the viewpoint of specific theoretical problems that needed an answer, and above all to some still mysterious expectations regarding its symbolic extra-scientific scope, which of course are the true reason for its success among specialists and not specialists.

Since we are facing a phenomenon of collective irrationality, subjective testimonies are not only useful but necessary, and it would be useful to have testimonies from those who have experienced within themselves the state of cognitive dissonance needed to adhere to the theory, and who later acquired consciousness of the contradiction they asserted. Unfortunately, those who like Dingle experienced this do not want to tell it, while those who like Essen did not experience it have no reasons for reticence, so they report their experience with simplicity, but in the latter case the testimony of the subjective aspect is much less interesting than would be in the first.

I tell my story because many readers who have tried and abandoned the study of the theory will recognize their experience. Without false modesty, I have to point out that I never experienced the cognitive dissonance state of acceptance, and instead I carefully observed the flattering of the sirens that attract everyone towards that condition of consciousness. By 1985 I had read some books and obtained some knowledge of the elements of special relativity without grasping the connection of the whole matter, and of course I thought that the lack of understanding was dependent on my limits and on the fact that I used too simple textbooks. I realized, however, that to understand the kinematic part of the theory it is sufficient to know simple things like speed and square root, so a feeling of discomfort remained imprinted in my memory together with the impression that this experience was different from other similar ones, when one gives up the work needed to understand something after an unsuccessful attempt: the single assertions I read all had a clear and simple meaning, while the complex they formed seemed meaningless to me. For the dynamic part I did not understand anything about the deductive chain that is used similarly in all textbooks to derive $E=mc^2$, but already I was struck by the observation that conclusions about mass and energy concepts were derived without there any relevant information in the premises. Everyone knows that syllogism: "all men are mortal, Socrates is a man, so all the fish live in the water" is not really a syllogism, although it is true that all the fish live in the water. Recalling this, I was wondering how it was possible to derive consequences about energy and mass concepts from premises that only contain kinematic information about the dimensions of space, time, and speed. I did not understand, but the doubt that the books I consulted were of bad quality was left in my mind.

More than thirty years later I made up my mind that the time had come to fill the gap, so I took the old textbooks and procured some more complex ones and for a few weeks I read with pleasure all the things we know, which are mathematically simple, and I realized without any difficulty Lorentz's transformations, the space-time interval, and Minkowski's geometry. In the meantime, I assumed that the phenomenon of dilatation was to be interpreted as apparent and symmetrical, because this had been explicitly taught to me by the first chapters of the textbooks I was reading. I had not yet come to the dynamic argument and the derivation of $E=mc^2$, but I was wondering: if all this is apparent, how can it ever be useful? We have to keep in mind that textbooks do not point out the problem of Maxwell's equations invariance, but they hint to it as a secondary issue, and so one of the consequences is that the theory may look like a fine game of formulas that is an end to itself, of which those who are endowed for math and not for physics can also be content. I read further confident that I would soon understand what the theorization of apparent dilatation was to serve, and I assumed that the twin paradox was a foolish tale foreign to the theory, without knowing that Bergson had had that idea too; obviously I assumed that the non-apparent and non-symmetrical slowdown of any minimal amount of a clock was part of the fable invented by the divulgation, like the macroscopic case of the twins. In the meantime, from hints to Lorentz's theory in the textbooks I realized that in that case the slowdown of the clocks would be real and not symmetrical: I did not understand the connection with Einstein's, but I was confident I would understand it by reading further carefully.

At this point, I came across the assertion that the slowdown of a clock is far from apparent, but it is real and can be verified locally when the clocks come to rest in the same place. I do not recall in which textbook I read it first, but I read and reread the pages of Resnick that we commented above in section 6.4, finding what we know: that according to the literature, time dilatation is a function of speed but is caused by acceleration, and the link between speed and acceleration is determined by the fact that we have defined a certain way of representing accelerated motion, Minkowski's geometry, which does not add any information to what we have, but only transforms available information. At this point I began to doubt I was facing some pervasive logical errors in the literature; doubts about the quality of relativistic literature had become such as to leave me more than perplexed, and I made the decision to carefully study ED1905, which I had previously thought to avoid because of the darkness of that text, which is too short to explain the matter exhaustively.

Meanwhile, I realized something else. Another rhetorical artefact of literature is to assert, but not to put in the necessary evidence, that the postulate of the constancy of the speed of light should be understood in the sense: "even by adopting the ballistic propagation model, all observers will measure c as the speed of a beam of light, whatever be their speed with respect to the source of the beam." The theory is expressed by affirming the abolition of the ether and the untenable Einstein's specific postulate, but readers must become aware by themselves of the exact meaning of the postulate and of the fact that it leads to the impossibility to draw any coherent geometry of motion. The literature assumes that propagation of electromagnetic waves occurs with respect to the emitting body, because this is taken for granted in all constructions to obtain the Lorentz gamma factor of the genre of the light clock; but never explicitly enunciates it. Then the literature asserts the absolute constancy of c, but never with the due emphasis to warn the readers that they must be prepared to accept an assertion that the relativistic literature also judges counterintuitive. As far as I know, it may well be that many readers go on reading the relativistic textbooks after having understood what the postulate seems to mean at first sight, namely that the speed of light is constant with respect to the emitting body: in this case, Einstein's deductions would not follow, but many readers may give up and abandon the attempt to understand relativity before focusing on these problems, and others may create their own interpretation of the theory and believe having understood what they have simplified and misunderstood completely. However, while I did not understand if we were talking about apparent and symmetrical or real and asymmetrical clocks delays, I continued trying to draw figures of the pattern of light propagation in the ballistic model, and continued not understanding how the constancy of c for observers in motion could be conceived. I realized that if these assertions are taken seriously and critically reflected upon instead of accepting them by mechanically repeating the sounds that express them, then they continue to lead into the Lorentz's hypothesis and not Einstein's. I did not know that in the framework of Lorentz the consistency of c is the logical and expected result (since the waves move with respect to the ether and not the source and since the observer's speed with respect to the ether is hidden by contraction), but while trying to attribute a sense to Einstein's postulate I continued to try to build a kind of stable tangle of electromagnetic waves with respect to which the contraction of lengths and durations of everything occurs; every time this representation came to my mind I discarded it because I realized that it violated the law of inertia, but what I did not know was that I was simply coming back to construct Lorentz's hypothesis by myself: such is the result of the ambiguity of the relativistic literature.

At this point, about a month had passed since the day I started reading the literature. Focusing on the issue of clocks rather than that of the sense of the postulate, I took in hand ED1905, and I found that the first chapters, the ones we commented above paragraph by paragraph, are a very brief display of what the literature exposes more extensively, but exactly correspond to the literature. Then I came across the expression: a clock "so gerichtet, daß sie die Zeit t' angibt", "adjusted so that it indicates the time t'" of which we have said in section 3.9. Reading this I concluded that in Einstein's text the sense of the theory was conventional, so there were no conceptual problems, and I was surprised (like Bergson) that the literature had turned a serious thing into a fairy tale, but it did not bother me too much, because the theory returned to appear perfectly logical. It remained to be understood why the convention to have set c absolutely constant even by assuming the ballistic propagation of light could be useful, and proceeding in reading I soon came to the fact that Einstein applied Lorentz's transformations as if the phenomenon of dilatation were real, although symmetrical (see above, section 3.11). Then I did not realize that I was facing the inability of Einstein's immature mind to control the coherence of the whole set of sentences he was saying, and I did not understand that in literature we face a phenomenon of dissonance and collective suggestion to elude dissonance and accept contradictions. The times were not ripe to draw these conclusions, and I hypothesized that in the specific problem of Maxwell's equations there was something special that allowed Lorentz's transformations to be applied to the interaction of the electric field with the magnetic field, as is suggested at first sight by Chapter 6 of ED1905. I tried to think if there could be such an explanation and posed the question to some experts, who did not give me an answer because they did not understand what I was talking about. This was useful because it indicated to me that I was out of the way: if in the specific case of Maxwell's equations there were a special reason authorizing Lorentz's transformations to apply to them, at least the most serious authors in the literature would say it, because that would be the key to resolve rationally the whole problem of ED1905 interpretation; everything else would be a fanciful extension.

About six weeks had passed since the beginning of the study, and I asked myself if it were possible that nobody has ever discussed this indeterminacy and this conflict between the conventional or apparent interpretation and the realist one? Surfing the internet, I found Essen's and Dingle's criticism. I procured their texts (which are not easy to find) and reading them I found the elements of the rational criticism of relativity which, after also reading Bergson's and Majorana's, I developed in this book. It should be noted that prior to the age of the Internet it would be difficult to find Essen's and Dingle's work: I discovered the existence of their essays in some unattractive web pages and consequently I looked for their original texts, but before the Internet era and poking only around library materials I do not know if one could find this basic bibliographic information, and how long the search would last.

In the four authors who rationally criticize relativity, neglecting some controversial marginal observation, I found a logical picture of things that was imposed with clarity and distinction beyond any doubt, and then I began to recapitulate the story in the way I have outlined it in this book. The difference between this book and the writings of the four authors is that this book is self-sufficient in the expression of the terms of the problem and can also be a starting point for readers that have just a vague idea of relativity, while the four authors all assume some more advanced knowledge of relativity by their readers. I read also Poincaré, and this reading was decisive in making me understand that Poincaré was not a timid and irresolute precursor of relativity as the relativistic literature suggests, but was a thinker who rationally considered in depth the consequences of the need for arbitrary and conventional assumptions for science. Einstein appropriated conventionalism immaturely and without becoming conscious of the inconsistency of the theory he elaborated, giving birth to an attempt that critical examination should immediately judge ingenious but failed. Finally, while comparing Lewis's 1908 paper with that of Lewis and Tolman of 1909, I ascertained that the doubt I had already realized thirty years ago was well-founded: no information about mass and energy can be extracted from purely kinematic premises. At most, one could conclude that a mass may undergo an apparent change in speed, as we have seen in Chapter 4 and Section 6.7, but it remains the usual misunderstanding between the apparent and real character of dilatation. For the remaining, the relationship of mass with speed requires dynamic experimental premises and it is a subject that requires the study of nuclear physics to be developed fully, and of course I have nothing to say about. I could not do anything else other than describe relativity as one could describe the astrological medicine that was in vogue in the Renaissance, and which Copernicus was also a doctor of: without being able to adhere to it, because I was inhibited by the cognition of its intrinsic inconsistency, but with the intent to make understandable the fortune of the theory in its time. The problem becomes to understand the exact meaning of the absurd theory that men have adhered to, and to understand from what human motivations they adhere to it.

In all this, I did not stop for a moment to wonder deeply about the show of cognitive dissonance, for which people belonging to and elite such as theoretical physicists seem unable to understand what they read in the face of a text that specialists could easily recognize to be a coarse application of conventionalism made by Einstein. But this leads us to the problem of the symbolic and extra-scientific value of the metaphor contained in the idea that time is not properly described by the classical framework: and this is an open problem.

Finally, even though I think it is obvious, I want to point out that I do not profess any opinion and any hypothesis in the face of the dilemma if there is something as stable as the ether and consequently Lorentz got it right, or if the propagation of light is simply ballistic, or whether there is a third model I cannot find anybody who has yet worked on it. I am only sure that a neoclassical generation of theoretical physicists full of life and strength sooner or later will reopen the problem by critically evaluating what is sensible and what is imaginary in twentieth century physics.

Conclusion

This is a blank page that present and future readers of this book will fill in with time. This book gives its readers a key that saves them the remarkable amount of work it takes to get out of the cognitive dissonance state in which one falls into when trying to reconcile with Einstein's specific postulate. Sooner or later someone will work out a critical review of the whole story and then decisive progress will emerge in countless fields, with theories that will remain faithful to logical solidity.

But in addition to the scientific problem of nature's behaviour there are many open issues concerning the attitude of men and problems of history and anthropology, which are revealed by the affair of relativity. Here are some in scattered order:

First of all, why was the cultural context of the time around the Great War so inclined to accept such a theory? What message, what metaphor, was implicit in the idea of multiple times and spaces? What anxiety received relief and what liberation was promised?

Then there is the problem of the psychological process of a false belief spreading in general. Now we know better than before that such processes have little to do with the average culture of the social environment where they happen. Here the world is upside down, and those who undergo the suggestion are an educated elite of scribes, not the world of the simple.

There is the problem of the specific rhetorical forms of literature that spread an impossible theory and its typical stratagems. All of us deceive first ourselves and then in good faith we deceive others by some constant structural elements: the hypertrophy of the discourse with respect to what is really pertinent, the alternation of logical arguments with messages now aggressive and now flattering towards the reader, and the appearance of depth generated by the brazen style of incoherent expression. In the minds of readers who do not understand is aroused a sense of inadequacy, and this is followed by indulgence and blindness to the logical mistakes of their own and of others to create an illusion of belonging to the group of privileged people who can understand. And here the writers share deeply the lack of self confidence in their readers, but they do their best to avoid showing it: aggression is part of the strategy to reduce the discomfort of cognitive dissonance.

There is the problem of leadership exerted by a person suffering some cognitive limit: the inability to control the complexity of what one says and the trust in all one's own fantasies, without being scrupulous of consistency, is a sign of a particular kind of personality, fully egocentric, where the satisfaction of libido silences categorically and radically the ability to judge objectively: we are all this way and we are all subject to the same temptations, but the degree of egocentrism is different among people. How does it happen that people who are incapable in the highest degree to realize that they are contradictory exert leadership and succeed in transforming the consequence of their cognitive deficit into social institutions?

There is a semiotic aspect of the problem. It was observed by Dingle that Einstein focuses on the page he is writing, that page, that chapter, without ever seeing the inconsistency of the whole. Suppose this corresponds to a subjective cognitive disorder. The result is that the contradiction is expressed boldly without attempting to legitimize it and the effect on the reader is the impression that the text contains revelations of unusual depth rather than unresolved contradictions. The impression is easily transformed into certainty and overall this is a phenomenon that does not depend on the subjectivity and intention of the author, but is produced by the cultural context through which the text is interpreted by readers.

The vulnerability of the twentieth century culture: knowing too many new things results in our world lacking the ability to synthesize. Institutions produced after the explosion of creativity in the early twentieth century are still uncertain and fragile in every issue of our lives.

Finally, what judgment should we give about the scientific community, or rather, to be right, about the elite theorists who speak a language so different from humble and modest technicians, experimental scientists, and technologists? Is it possible that nobody is interested in knowing, for example, whether the emission theory could replace that of the ether, and so perhaps in using the electronic technologies accumulated up to the present to go to the bottom of the problem? The fact that maintaining consensus about relativity interests this privileged world far more than questioning nature is the most amazing revelation we receive when we acquire the exact cognition of the meaning of special relativity. Anti-classical theoretical physics of the twentieth century is a world of its own, speaking a language of its own and has little to do with experimental science and technology in every field. How many critical reviews are there to be done about the results of anti-classical theoretical physics? Is the credulity towards the inconsistent myth of relativity the only sin, or is it the original one?

Bibliography

This bibliography has the sole purpose of identifying the texts mentioned in the course of the discussion. It is easy to imagine that a complete bibliography of special relativity, or even what was written on Einstein, would fill volumes. It does not appear that there is a bibliography of rational critics of relativity, but the reading of the writings by Dingle, Essen, and Majorana, each of which mentions numerous reliable authors, would allow one to gather a list of critical positions expressed by non-compromised authors with the world of pseudoscience. In this way, it would not be difficult to compose a complete picture of rational criticisms expressed until the 1970s, while for the times closest to us the problem is more difficult for the dizzying growth of fanciful and pseudoscientific publications and the absence of new voices which support rational criticism.

For those who want to undertake research on the issue of relativity in the critical perspective, I suggest that the first book to read, containing a lot of data and a more extensive bibliography, might be that of McCausland's (2011), an electrical engineer who was in contact with Dingle and continued his work, describing his own experience in *A Scientific Adventure*. There are many useful indications from which to start the study of the history of relativity also in anthropological terms.

AA.VV. e Einstein, "La théorie de la relativité", in *Bulletin de la Société française de philosophie*, 1922, pp. 349-370. Report of the meeting of Einstein with Bergson and other (Xavier Léon, Langevin, Hadamard, Cartan, Painlevé, Paul Lévy, Jean Becquerel, Brunschvicg, Le Roy, Bergson, Meyerson, Piéron). The text is available on line, but with many OCR errors: http://www.sofrphilo.fr/activites-scientifiques-de-la-sfp/conferences/grandes-conferences-en-telechargement/

AA.VV., *Hundert Autoren gegen Einstein*, Leipzig, R. Voigtländer Verlag, 1931.

Bergson, Henri, *Essai sur les données immédiates de la conscience*, 1889. In English this essay was translated with the title *Time and Free Will*.

Bergson, Henri, *Durée et Simultanéité. À propos de la théorie d'Einstein*, 1922. The second edition, of 1923, is more complete because it contains three important appendices written after the meeting between Bergson and Einstein.

Breithaupt, Jim, *Understand Physics: Teach Yourself*, McGraw-Hill, 2002 e 2010.

Canales, Jimena, *The Physicist and the Philosopher: Einstein, Bergson and the Debate that Changed Our Understanding of Time*, Princeton 2015.

Chevalier, Jacques, *Entretiens avec Bergson*, Paris, Plon, 1959.

Dingle Herbert, *Relativity for all*, Boston, 1922.

Dingle Herbert, *The Special Theory of Relativity*, London, Methuen & Co, 1940, 1961.

Dingle Herbert, *Introduction* to the English translation of Bergson's *Duration and Simultaneity*, Bobbs-Merrill, 1965.

Dingle Herbert, *Science at the Crossroads*, London, Martin Brian & O'Keeffe, 1972.

Einstein, Photoelectric effect 1905, "Über einen die Erzeugung und Verwandlung des Lichtes betreffenden heuristischen Gesichtspunkt", *Annalen der Physik*, 17, 132-148 ("On a heuristic point of view concerning the production and transformation of light").

Einstein, ED1905, "Zur Elektrodynamik bewegter Körper", *Annalen der Physik* 17, 891–921 ("On the electrodynamics of moving bodies") about special relativity.

Einstein, Inertia 1905, "Ist die Trägheit eines Körpers von seinem Energieinhalt abhängig?", *Annalen der Physik* 18, 639–641 ("Does the inertia of a body depend upon its energy content?") on the equivalence of mass and energy.

Einstein, January 1911, "Die Relativitäts-Theorie", Naturforschende Gesellschaft in Zürich. Vierteljahrsschrift 56. Part 2, Sitzungsberichte (1911): IX. (The Theory of Relativity").

http://einsteinpapers.press.princeton.edu/ document 17 of 1911

Einstein, May 1911, "Zum Ehrenfestschen Paradoxon. Bemerkung zu V. Varičaks Aufsatz". Von A. Einstein. Mai 1911. (Eingegangen

18. Mai 1911), *Physikalische Zeitschrift* 12 (1911): 509-510 (On the Ehrehfest Paradox. Comment on V. Varičaks's Paper).

http://einsteinpapers.press.princeton.edu/ document 22 of 1911.

Einstein, *Relativity: The Special and the General Theory*, Original title: *Über die spezielle und allgemeine Relativitätstheorie (gemeinverständlich)*, 1916. Many editions of this book are available in English and other languages.

Einstein, November 1918, "Dialog über Einwände gegen die Relativitätstheorie", published on 29 November 1918 in *Die Naturwissenschaften*, 6, (1918) pp. 697–702. ("Dialogue about Objections to the Theory of Relativity").

http://einsteinpapers.press.princeton.edu/ document 13 of 1918.

Einstein, Albert e Infeld, Leopold, *The Evolution of Physics: The Growth of Ideas from Early Concepts to Relativity and Quanta*, 1938.

Essen, Louis, *The Special Theory of Relativity: A Critical Analysis*, Clarendon Press, Oxford, 1971.

Essen, Ray, *The Birth of Atomic Time*, 2015.

Fabri, Elio, *Insegnare relatività nel XXI secolo*, Associazione per l'Insegnamento della Fisica, 2005.

Ferraro, Rafael, *Einstein's Space-Time. An Introduction to Special and General Relativity*, Springer, 2007.

Festinger, Leon, *When Prophecy Fails*, University of Minnesota Press, 1956.

Festinger, Leon, *A Theory of Cognitive Dissonance*, California, Stanford University Press 1957.

Isaacson, Walter, *Einstein. His Life and Universe*, Simon & Schuster, 2007.

Langevin, Paul, Bologna Lecture of 1911 in *Scientia* n. 10, 1911, pp. 31-54. Can be read on line at http://amshistorica.unibo.it/scientia

Lewis, Gilbert, 1908, "A Revision of the Fundamental Laws of matter and Energy", *The London, Edinburgh, and Dublin Philosophical Magazine and Journal of Science*, November 1908, 6, 16, (95) pp. 705-717.

Lewis, Gilbert e Tolman, Richard, 1909, "The Principle of Relativity, and Non-Newtonian Mechanics", *Proceedings of the*

American Academy of Arts and Sciences, Vol. 44, No. 25 (Jul., 1909), pp. 711-724.

Lorentz, Hendrik, 1916, *The Theory of Electrons*, Teubners, Lipsia. The first edition is of 1909.

Majorana, Quirino, *Le teorie di Alberto Einstein*, discorso tenuto dal Prof. Quirino Majorana all'Accademia delle Scienze in Bologna, 9 Dicembre 1951.

Majorana, Quirino, "Cinquanta anni di relatività e sulla soglia di una nuova visione della fisica", *Sophia*, II, 1956.

McCausland, Ian, *A Scientific Adventure*, Roy Keys, Montreal, 2011.

Miller, Arthur I., 2001, *Space, Time and the Beauty That Causes Havoc*, Basic Books.

Ortega y Gasset, José, *El tema de nuestro tiempo*, 1923.

Pais, Abraham, *'Subtle is the Lord...' The Science and the Life of Albert Einstein*, Oxford University Press, 1982.

Pesce, Domenico e Pozzi, Lorenzo, *Primi elementi di logica formale antica e moderna*, Le Monnier, 1971.

Poincaré Henri, *La Science et l'Hypothèse*, 1902.

Poincaré Henri, "L'état actuel et l'avenir de la physique mathématique", Conférence au Congrès international des arts et des sciences, Saint-Louis, Missouri, 24 septembre 1904, in *La Revue des idées*, November 1904, 801-818. And also, *Bulletin des sciences mathématiques* 28, 1904 (décembre), 302-324. And also in Poincaré [1905], Chapters 7, 8 e 9. Translated into English by George Bruce Halsted, it appears as an appendix to the volume *Science And Hypothesis*, New York, The Science Press, 1905.

Poincaré Henri, *La valeur de la science*, Flammarion, Paris, 1905.

Resnick, Robert, *Introduction to Special Relativity*, John Wiley, 1958. PDF available at the site www.archive.org

Rohrlich, Fritz, "An elementary derivation of $E=mc^2$", in *American Journal of Physics*, vol. 58, n° 4, April 1990, p. 348.

Ronchi, Vasco, *Storia della luce*, Laterza 1983.

Trefil, James, *The Routledge Guidebook to Einstein's Relativity*, Routledge, 2015.

Varičak, Vladimir, "Zum Ehrenfestschen Paradoxon", *Physikalische Zeitschrift*, 12 (1911), pp. 169-170.

Back cover

Everyone knows that Einstein's special relativity contains a theory of time measurements, which are no longer conceived as absolute, but are related to the state of motion of the clock and to the point of view of the observer, and the same happens to space measurements. Everyone also knows that the theory contains the deduction that a small material mass can be converted into a huge amount of energy according to a precise quantitative relationship.

But many who have tried to study the theory have failed to understand it; yet, to fully understand the part of Einstein's theory about time and space measurements, readers just need to know what speed and square root are, and to obtain a simplified but clear idea of the part regarding the concepts of mass and energy they need just to remember elementary high-school physics. Apparently something is missing in all the many books that describe relativity in a simple or higher level.

This book is written in a different way from any other. A rigorous but clear exposition will show all readers, provided they know what speed and square root are, that they can understand fully and perfectly the space-time theory and can judge it with their own intelligence. In addition, readers will have a clear idea of the equivalence between mass and energy and its logical relationship with space-time theory.

This book was written for beginners and for perplexed people who have unsuccessfully attempted to study special relativity: both will understand the exact meaning of the famous and difficult essay in which Einstein expounded the theory in 1905, which is examined word by word in this book. And all readers will have a clearer idea of the relevance of relativity for the twentieth (and twenty-first) century culture.

Alberto Palazzi

A scholar of history and philosophy and designer of computer algorithms for problems of higher complexity, the author of this book brings together the skills necessary to analyze the scientific significance of relativity and to reconstruct the historical and anthropological context in which it was born and became an institution of our time.